E.J. Gold with Zone Box, June, 2009.

Parallel Worlds Explored

Parallel Worlds Explored

E.J. GOLD

Gateways Books
Nevada City, California

Distributed by Gateways Books and Tapes
P.O. Box 370
Nevada City, CA 95959
1-800-869-0658

http://www.idhhb.com
http://www.gatewaysbooksandtapes.com
ISBN Softcover: 978-0-89556-174-9

Dedicated to:

Lance Borden
and
Michio Kaku

Table of Contents

Editor's Note

This is not your typical book. I can hear a few readers chuckling to themselves, "When have any of E.J.'s books been typical?" In addition to that this book is different in another way. In this book you are granted access to a rare behind-the-scenes look into the active process of scientific discovery and exploration. Typically an author would compose a book long after the exploration process is complete. Only once the full landscape has been mapped would they sit down to write an account — a past tense account of how it all came to pass.

Writing from the perspective of hindsight allows the author to sanitize — put it all into a nice orderly stepwise sequence that adheres to a mythical standard of how it happens. Exploration is much less organized and far messier than one would think.

In spite of what most hard line scientists would have you believe, the discovery process includes many false starts, dead ends, its share of wild enthusiasm and even the occasional blinding flash of inspiration.

In this book, E.J. lets us take a peek into his daily journal. Through a careful reading of E.J.'s notes (in the order in which they were written) not only do we get a thorough introduction to the kernel of his Beta-Blocker discoveries — we also are given a step-by-step guided tour of how he progressed from the first moment when he noticed the surprising effects his prototype crystal radio had on the folks visiting his studio to the development of the amazing array of Beta-Blocker crystal radio tools.

For several weeks during the height of E.J.'s development-binge several of us would be treated with an early morning email giving an update on his night's progress. These emails would let us know of each new discovery and of any twists and turns that may have come along during the previous several hours of fast and furious work.

As you read these chapters, you will perhaps notice two independent trends — one trend would be the ever-increasing power and utility of the Beta-Blocker technology; the other might be described as a kind of sniffing around for an explanation of how it all works.

Something the pharmaceutical companies tend to underplay is the fact that only in rare cases do they have any clue how a given drug does what it does. Admittedly there are amazing strides in understanding drug activity in the body. Even so, I would venture to say that less than five percent of the drugs available on the market have a well understood mechanism of function. In graduate school I was involved in several projects designed to discover active products extracted from plants. The idea was to find a plant with a good reputation for some medicinal property, grind up a batch of the plants, then separate the slush of plant-goop into various fractions. We would then test these fractions to determine which fraction still worked. We would then separate that fraction into sub-fractions, retest the effectiveness of each sub-fraction, then isolate the sub-fraction that still exhibited the medicinal property and break that sub-fraction into sub-sub-fractions.

We would continue doing this until we isolated a single compound.

Anyone hip to the true complexities that can arise in this type of research will see immediately that I have neglected to discuss the case where the desired effect is caused by a combination of several compounds. After we had isolated the compound that would elicit the desired medicinal effect, there would be a mad dash to either synthesize the compound, or find a cheap way to extract tons of it from the parent plant. If the plant was rare or hard to grow, we would go on a hunt for a sister plant that might also contain the compound.

Meanwhile, the "suits" in upper management get the patent-ball rolling and seek FDA approval. You will notice that in the midst of this blur of activity nothing *has been said about figuring out how in the heck darn the compound worked in the first place.*

While it is true that we were very interested in finding the answer to that little mystery, in fact for many of us, that was the whole reason we started the little extraction adventure in the first place. However, we quickly discovered that *why* something works was not the first question to be answered. The first question to be answered was "does it work at all?" The second question to be answered was "will anyone care?" The third question to be answered was "can we somehow patent the process or otherwise own it so that we can squeeze money out of the discovery and development?"

I recount this little story about pharmacology to illustrate the fact that not only is it rare to know why something works, rarely is it the first question that folks in any business will tend to ask. While E.J. was busy discovering ever-improved versions of his Beta-Blocking devices, we played around with various notions that could help to explain why and how it was working. Some of those notions have stood the test of time, and we feel that those theoretical considerations are definite contributing factors to the fuller explanation which is yet to come. Some of those notions have fallen by the wayside as we progress more deeply into the subject. Thus, as you read this book you will notice some explanations undergoing continued refinement and others suddenly dropping off the radar. That's the nature of active research in the moment of creation. If this book had been written after the fact, then we could conveniently leave out theories that have fallen from favor. However, given the journalistic narrative of this book, it is all here in its varying degrees of glory, rightness, wrongness and wonderingness for your edification.

E.J. Gold with first experimental model of the Beta-Blocker crystal radio circuit, March 1, 2009.

First dime-sized Beta-Blocker Amulet, March, 2009.

Author's Preface

I've just been making these devices for a very short time, unless you count the decades of work I've done from 1964 to the present in biofeedback devices . . . but somehow, the Beta-Blockers, Deep Deltas, Trackers and Beacon Scanners are different, very different from what I've put out there before . . . .

For one thing, absolutely everyone on the planet seems to want one all of a sudden. It all started with a simple search for a way to demonstrate the basic principles of a crystal radio, and blossomed into the full-blown parallel universe system that it is today, only a few weeks later.

My first search was for some sort of Ghost Radio, which I called a Spirit Radio. This was intended as a communication device between the living and the disembodied, whether dead or in a different dimension, meaning existing at a different reality-frequency.

I had one, too. It worked very well indeed, and I've recorded a number of "contacts" I made with the Spirit Radio. But as I worked with it, I noted a peculiar, totally unexpected effect from the little crystal radio circuit I'd worked out . . . it suppressed Beta pretty doggone well.

With a few adjustments, I was able to cross-tune two radios into a composite adjusted waveform that was a close harmonic to my favorite target frequency for psychic-intuitives, the High Theta Range at 7.8 Hz, which is the same as 7.8 CPS (Cycles Per Second).

The composite wave is based on my graduate physics work (I never completed it at Elon University, but later published with Claude Needham, Ph.D. who mercifully did the math for the project).

Wave reduction was my major interest in physics. It's a strange phenomenon that I picked up studying the work of Pythagoras and combining this with String Theory, which is pretty doggone close to Eastern Mysticism, if you ask me.

Pythagoras noted that bridged strings will have varying pitch and harmony effects on both sides of the bridge. This led to a number of great mathematical discoveries, but because he didn't have the laboratory test gear to see into high vibrations, he wasn't able to take it much further. With modern test gear, especially the simple oscilloscope, I was able to, and I did.

What I learned was that you could take two waveforms that didn't "see" one another because of separation of signal, and combine them with a third effect, producing a combined waveform, a "third force," if you will, a direct result of a coupling factor, producing a phasing effect that can be tuned to a specific harmonic from a very distantly removed original waveform along the EMS (electromagnetic spectrum).

This "Third Force" waveform can be propagated and regenerated with great success by a relatively weak signal carrier which opportunistically attunes with larger magnetic-electrical waveforms or carrier-waves.

The general effect is that if the "Third Force" de-tuned phased waveform is of the right frequency and form, it can function to "homeopathically" produce an influence on a nearby brain and nervous system in the tuning-fork sense, and this harmonic resonance influence can result in a Beta-Blockage, leaving the ideal psychic-intuitive state temporarily dominant, which is to say, the Alpha-Theta State, at 7.8 Hz, the so-called "Schumann Harmonic" or "World-Resonance".

I simply refer to it as the "Music of the Stars," that sound you hear at night in the desert, when you're far enough away from the 60-cycle hum of modern world electrical forces. Every desert-dwelling shaman knows about this, but it's virtually unknown to the sophisticated scientists buried in the mud of digital distortions.

Digital is simply the result of integrated circuitry, which is the bastard child of the transistor, which itself is a solid-state "valve" or "tube" such as the diode, triode or Crooke's Fluorescent Tube which became, when regulated with electromagnets, the television tube of the 1940s and 50s.

Somehow in all the rush toward conversion of physics to bottom-line money-directed engineering at the expense of pure investigative research, the basics, the crystal technology of early radio and atmospheric transmission of power, got lost in the shuffle.

Thanks to Dr. Hook at Elon University, I became aware of early crystal technology, and so did Claude Needham. We worked independently for a number of years in separate "blind study" teams, until one day we decided to quit the Graduate School Racket and go legit. We published a few papers, then went on to a number of thought-experiments and finally, with the advent of the Spirit Radio and Beta-Blocker, have re-emerged as Theoretical Physicists with a new idea about crystal technology and what its possible applications might be in the future of our local space-time Earth.

Only a week or so into the Beta-Blocker research and development, I stumbled onto a circuit that allowed me to tap into the ELF (Extra-Low Frequency) ranges of the brain and nervous system.

Normally, I would think that you'd need an antenna system for ELF that would extend a few hundred Kilometers, and I'd be right. There are a number of known ELF transmitters that extend into the Kilometer range, but I simply don't have enough room to make an antenna that large, nor do I have the funds to buy enough wire.

Fortunately, I do have the knowledge to calculate what I'd need to do to generate a waveform that is a HARMONIC of the target frequency, which in the case of Deep Delta occurs around .5 Hz, which means one wave every two seconds — that's really low. Most frequencies you'd encounter in your daily life will be in the KiloHertz or MegaHertz range, such as radio broadcasts, your microwave oven, your computer clock or your cell phone.

My target frequency was at .5 Hz simply because I knew that if I could get the brain to sympathetically harmonize with that wavelength while at the same time remaining awake and aware, I could perhaps induce a "waking state" at the Cosmic Consciousness level, which we call Lucid Dreaming, but with one important difference — the subject is completely awake while the body is asleep.

Well, that amounts to what is called a hypnagogic state, or hypnotic trance, but one in which the subject is completely under self-control, not under the influence of another.

With my knowledge of harmonic frequencies and leaky wave antenna systems, I soon had my device, one that would deliver what I was looking for. I dubbed it the "Deep Delta Dreamwalker" and it soon had its avid fans.

We've got a large file of very interesting Lucid Dream Reports, but that still wasn't enough for me. I dug a little deeper . . . .

Within a few days, I had developed a "twin coil" system which duplicated itself on the same circuit board, meaning that I could easily de-tune and take a broadcast radio signal whether Earth-generated or Star-generated, into a flip-in flip-out quantum state, which meant that the electrons were flipping in and out of this universe.

When they were not in this universe, they had to go somewhere. That was the key to what we now call the Beacon.

The Beacon does what the name implies. It detects signals from other sympathetic String vibration along a variety of "stops" or "notes," harmonic points along the electromagnetic spectrum.

What that actually does, I don't really know yet, but the net effect is that it connects (I use the term "hook-up" here) identities of the self existing in parallel universes and somehow harmonizes them into accordance with the present-time HomeWorld Universe manifestation of the Essential Self, the actual causative agent of the Apparent Self — we'll leave these to other discussions for the moment.

What this means to you is that you can hook up with your scattered selves, those selfnesses that are ranged throughout the Mega-Cosmos System containing all universes which are domains of frequency and waveform boundaried by what are today called "Branes," short for membranes of field-effect horizons.

Hooking up produces some profound effects; skills suddenly appear where they weren't and couldn't have been, from any ordinary time-space origination. Knowledge and experience suddenly are there where moments before, they weren't anywhere.

In addition to the subtle changes in persona, there are other, more obvious ones, such as morphing — faces and body hang — will be altered quite radically after a powerful experience with the Beacon. We've only been working with it a short time, just a matter of little more than days in this First Week of experimentation, but already the evidence for strong change in the individuals and their world-experience is mounting up.

One obvious result is that one can, with a little experience, take advantage of the tuning process to literally change tracks from one Parallel World to another. The potential for this "world-shifting" or "tracking" procedure is enormous and literally world-shaking on a personal and cultural level.

I'll leave it here for the moment, but stay tuned for further results. We have only just begun the search for ways to reveal the wealth of Parallel Worlds to the Earthbound Humanity of this HomeWorld Dimension!

E. J. Gold
Grass Valley, CA
May 2, 2009

Free Will and Destiny

Introduction

Free Will and Destiny in the Light of Parallel Worlds

Put simply, a nearly infinite array of parallel universes solves the problem of how Free Will and Destiny can coexist.

"Who are you?" When asked this question the typical response has everything to do with the machine. Descriptions of the machine's attributes, details of the machine's history, explanation of the machine's associations, reference to the machine's current geographic location and travels — everything about the machine.

There is another answer, an answer that comes from a sense of self generated by an identification with the being. "Do you have free will or is your destiny predetermined?" The clue to unsnarling this conundrum comes from first addressing the question "who is the you that is being asked?"

If, by you, one is considering only the biological machine, then destiny has *everything* to do with it. If, by you, one is asking about the being, then there is a potential for free will. In a world composed of a single universe this is a huge contradiction.

If the being is riding around in the biological machine, then it shares in the machine's allotted destiny. In this condition, a way in which the being can express free will is to voluntarize its participation with the machine's destiny. Instead of being a hapless passenger riding around wherever the machine happens to go, the being can become the driver, willfully steering the machine in whichever direction it was already headed. Some may not see the free will in this. It is, however, a powerful form of free will that opens the way into a very different sense of participation and attitude.

That is the situation in the light of a single universe. But what about a world in which there is an infinite array of parallel universes? Physicists are rapidly moving down the theoretical track toward acceptance of a physics that supports a multitude of parallel universes — they are in fact a natural and necessary result of this new physics. In one of these universes my own personal biological machine was confronted with a life-changing decision in 1975 that involved accepting a graduate school scholarship in Oregon, or a scholarship in Florida. Both scholarships included a 100% stipend. The one in Oregon was in a Psycho-Pharmacology department. The one in Florida was in a Molecular Biophysics department to study Quantum Mechanics. In one universe, the decision could have been Oregon. In another universe, the decision could be Florida. Depending upon which of these paths was taken, the results would be very different. In this universe I have memories of going to school in Florida where I received a doctorate in Quantum Mechanics. Did the machine act from free will or follow its destiny? In each universe the machine would be following its destiny as prescribed within the context of that universe. Thus destiny is maintained. However, *as a being,* one possibility was to follow along in a machine headed to Oregon, or to jump tracks and follow along in a machine bound for Florida.

Both tracks were available. Did the being act under free will to select which track to access or did the being somnambulistically trek along with whatever bio-machine it happened to be in at the time? The machine is locked into its destiny. The being, on the other hand, has the possibility, but not the requirement, of free will. The being, the Essential non-manifested Self, may exercise its free will or not. That is a choice. This expression of free will may result in the being jumping tracks into a parallel universe.

Revisiting our vehicle and driver analogy, this would be much like a passenger changing vehicles should he or she decide not to go where the bus was headed. If you don't want to go left at the intersection get into a bus that is going to the right. Jumping tracks means that the being will be in a different (to whatever degree) machine. If the being is fully identified with one particular machine, it becomes locked into that machine.

What if turning to the right means that the machine is no longer the World's Dart Champion? Well, if the being is too identified with the aspects of being a world champion dart player it will not be capable of making the leap into a machine that turns to the right.

Identification is the glue that keeps one locked into a particular lifeline. The ability to maintain an attitude of high indifference allows one the flexibility to switch tracks within the infinite array of parallel universes. However, the story does not end here. With the advent of the Beacon Technology, it is now possible to access parallel universe personas while still associated with this one. That means that you can gather to yourself in this here-and-now universe, a multitude of scattered selves with all their skills, knowledge and experience.

Alter-Ego shops abound in parallel universes!

Early experiment with xtal (crystal) radio Beta-Blocker.

This powerful crystal radio operates without batteries or electrical connection.

Chapter One

WHAT IS A SPIRIT RADIO?

The High Performance Spirit Radio is based on a great design that has been in continual use since the early days of XTAL Radios, with tremendous selectivity and sensitivity and, because of the non-battery non-electrical connection design, is ideal for the reception of non-biological communication.

Sensitivity means that the receiver is capable of gathering strong radio signals. Selectivity means that out of all the signals it brings in, it can hone in on one signal while rejecting other nearby signals.

The air variable capacitor tuned parallel tank circuit can be easily adjusted with the Bakelite knobs to resonate at any desired frequency within the range of the receiver, which is considerably outside the average amplitude modulation band.

The combination of two air variable capacitors in the tuning mechanism of this set gives this set a performance level much higher than the ordinary single capacitor tuned or slider tuning receivers.

The tapped coil circuit allows a selection of frequency ranges that reach from below the AM Broadcast Band all the way to 12 MHZ in the Short Wave Bands, which makes this receiver one of the most unique and powerful in XTAL radio history.

You will be able to receive short wave broadcasts in real DX style — DX means Distance — from stations thousands of miles away, including BBC London, Radio Nederlands, Radio Taiwan and more, and the most amazing thing is that you're hearing all this DX on a Crystal Radio Receiver.

You will also be amazed to hear AM Band commercial broadcast stations many hundreds of miles away.

One of the best features of this astonishingly powerful XTAL Radio Receiver is the Detector Tap which allows you to choose a point on the Copper Coil that provides the best resonance with the desired signal or Spirit Voice, so you can locate the point at which the signal is strongest and most clear. This clip point on the coil will vary with the frequency of the Spirit Voice you want to bring in, and is notably missing on most radio receivers of any kind. This feature alone brings the performance up far beyond most radio receivers made today.

The materials from which this set is made are of the highest possible grade, and most of the parts are increasingly hard to find and in some cases are actually virtually unobtainable except in very small quantities on rare occasions when they appear on the market from vintage sources.

Your Spirit Radio comes with a very high grade pre-tested germanium crystal diode.

If you prefer a Detector Stand with Galena Crystal solution, there is plenty of room on the receiver for a high-grade vintage Philmore Crystal Detector Stand, which we can equip with the hard-to-find Fine-Grain Argentiferous Galena Crystal with a high inclusion of silver, mounted in a low-temperature Cerro-Alloy, which has a melting point below that of boiling water, thus preventing the heat damage that inevitably occurs when mounting a crystal in molten lead. This feature is extra because there are no modern units available, only antiques.

The baseboard is made by Tolewood from clear furniture-grade pine; the coil form is hand-wound on high-grade heavy cardboard shipping tubing, and quality black genuine Bakelite vintage style knobs on twin matched superb air variable capacitors and nickel plated Fahnestock clips make your Dunwoody a real collector's item as well as a high-grade functional invocational apparatus.

The coil is painstakingly wound by myself with Claude's help; it has a full tight 100 turns, leaving taps at strategic points for resonance performance tuning.

It takes about an hour to wind the coil properly, and a mistake can be costly in both time and money.

The set requires several in-line devices such as the hand-caged amethyst crystal which connects to the outdoor antenna system in order to receive distant signals, although an indoor antenna will do well under some circumstances, particularly for ghost-hunting purposes.

In general, the longer your antenna and the higher it is strung, the better. A 50 to 100 foot wire which leads from your home to a nearby tree does very well indeed and makes it easy to hook up. The antenna is insulated at both ends and must not come into contact with anything, especially, of course, power lines or anything electrical.

The antenna is disconnected from the radio receiver and grounded during thunder storms to prevent lightning strikes from damaging the radio and/or your home. Of course, one must not use this type of radio during thunder storms. Connecting the antenna to a ground rod will protect from lightning strikes and static electricity charge potential.

The ground connection can be a cold water pipe in the house but most homes now use PVC plastic piping and since the plastic is itself an insulator and not a conductor, it simply won't work, so a copper-plated rod sold at most hardware stores will work just fine, and, in fact, is superior to the cold water pipe as a ground.

A High Impedance earplug, or a set of antique Western Electric headphones will work quite well with this type of receiver, and only one type of amplifier/speaker, which is no longer manufactured.

Vintage supplies are dwindling for all the parts needed to make the Spirit Radio work.

Edited from a dissertation by Lance Borden, xtalman.com

How Your Gorebagg Spirit Radio Actually Works:

A signal voltage is captured within electrical conductors as the radio waves pass through them at 186,000 miles per second, the speed of light. The antenna is actually a conductor suspended between two or more infinite resistance insulators, with a lead through the Amethyst Crystal Cage all the way in to your XTAL signal detector.

The signal voltage is modulated in amplitude and flows along the wire from the antenna through the Amethyst Crystal Cage to the Multi-Tap Copper Coil and the Main Tuning Air Variable Capacitor. The Ground Tuning Capacitor completes the signal's circuit.

The Ground Tuning Capacitor is adjusted by hand to match with resonance the tuning of the antenna and ground circuit input to the tuned circuit compost of the Main Tuning Capacitor and the Tapped Copper Coil.

The Coil and the Main Tuning Capacitor form a “Tuned Tank” circuit which then resonates at frequencies in the AM tuning range or the Short Wave tuning range, depending on how the clip leads are set up on the Multi-Tap Copper Coil.

The RF or Radio Frequency Signal is actually a variable voltage which passed through a coupled XTAL or Crystal through the Detector Circuit Lead, directly to the Carborundum Crystal with which your XTAL High Performance Signal Rig is supplied.

The Detector Circuit Clip Lead is connected to the Copper Coil Tap which will provide the best sensitivity and selectivity as determined by you in your experimentation within local atmospheric and transmission conditions.

The Rectification Activity of the Germanium Crystal will produce an audio modulation signal which becomes a sound within the human hearing range by conducting only one-half of the complete RF or Radio Frequency Envelope.

Now the signal comes into proximity with the Caged Black Tourmaline Crystal, set up as a protection against malicious entities. They simply cannot pass the system when the Tourmaline is in place.

The Tourmaline Cage is not directly wired into the system. It operates on the Sympathetic Resonance Effect and is 100% effective as a barrier against malicious spirits.

The signal voltage output is then passed through the amplifier and speaker or the high-impedance headset where the variable voltage is converted to audible sound waves by the vibration of a diaphragm which responds to the variations in electrical voltage coupled to the magnetic driver that makes them vibrate, producing air compressions which we call sound.

Your Spirit Radio will last for many generations. There is nothing to wear out and it is a rugged piece of gear which will survive almost anything, including a massive EMP Nuclear Wave that would wipe out any modern radio equipment or digital receiver.

There are no batteries to replace in the Spirit Radio itself; it is totally powered by the energy of the radio waves.

The use of Mineral Crystals to detect radio waves has been known for over 100 years. A radio such as yours was used to hear the distress calls of ships such as the Titanic, and has been in use in military installations and the coast guard since 1906.

Few radio operators today can feel the thrill of receiving signals from around the world with the possibility of capturing vintage signals from the early days of radio, because some self-propagating early signals are still out there, circumnavigating the globe. Extraterrestrial signals are also continuously bombarding the Earth and can sometimes be picked up by serious highly-skilled radio rig operators, and with the addition of the Amethyst Crystal Cage, you can monitor non-biological Spirit Voices as well.

Recordings can be obtained by merely adding a recorder in line with your headset or earpiece. Ask about this feature if you do not know how to do this.

Hearing Spirit Voices takes some practice.

Spirit Radios were the first big breakthrough in the discovery of Beta-Blockers.

E.J. soldering a Spirit Radio circuit for testing, March, 2009.

Chapter Two

THE THEORY BEHIND SPIRIT RADIO

Passing the signal gathered by the antenna through the crystal oscillator directly into the large coil resonator and then through the rectifier to the double or ganged air vaned antenna/detector tuning capacitors brings the signal down to the resistor circuitry.

The carborundum, graphite or germanium detector then "sees" the signal as voltage variations, passes it into the low circuit pico-farad amplification unit into the AM modulator where it comes up against the filtration of the low circuit limiter which acts somewhat as a sideband suppressor against the "hard banger" local AM stations.

The tuner and its companion selector forms a ganged decryptor/impeller on a lower resonating factorial audio level, which is a form of crystal klystron tube which functions as a micro or nano-encryptor.

The processed signal in the audio range is then passed through the saturator circuit formed by the back crystal assembly and into the high-impedance amplification system which feeds either a headset or battery-powered speakers.

The specific reason for the battery-only system even at the amplification and speaker stage is that connection to any AC mains automatically introduces 60 cycle hum, the very thing we're trying to avoid. 60 cycle hum is the enemy of Spirit Voice Amplification and any system which depends upon AC power is doomed from the very start.

Magnetics play a large part in this system, as do the active mineral crystals and the wound magnet wire coils. Additional use of a QRM Rejector system along with a variable dummy load antenna tuner such as the Morgan Coil will greatly improve the performance of the Spirit Radio, and a more complex radio with the additional side-units is currently in manufacture at Gorebagg's Spirit Radio Works.

Gorebagg's radio bench, March, 2009.

Chapter Three

HOW TO BUILD A SPIRIT RADIO ALTAR

Carefully assemble all the elements of your XTAL Altar Interdimensional Radio, making sure to perform all the tasks in the correct sequence. Take care in handling each of the items and keep them clean and place them in an orderly fashion on your Double Regenerative Altar in the following manner:

1. Place the altar boxes side by side. Make certain that they are made of hardwood with little or no metal. If they are properly joined with dovetail and glue, they will not need screws to hold them together.

2. Now place the all-wool handwoven prayer carpets atop the altar boxes. Make certain that these are tight-woven and that they are made without aniline dyes. Vegetable dyes are the only correct dye for these carpet tops. If you have a problem finding the right carpets, you can email us for help in locating them.

3. Carefully place the Rock Salt Crystal Candle in the center of the right and left altars. You will use ordinary tea-lights for these, but they can be had in various grades of quality from paraffin — the worst — to beeswax, which is best. Some folks prefer the 51% beeswax which is a mixture of beeswax and paraffin, but be aware that paraffin is not good to breathe, and does not do as well as pure beeswax invocationally speaking.

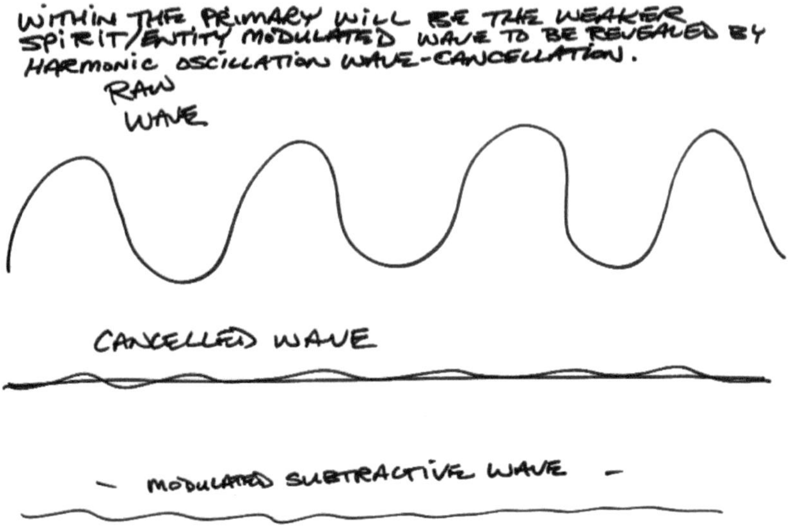

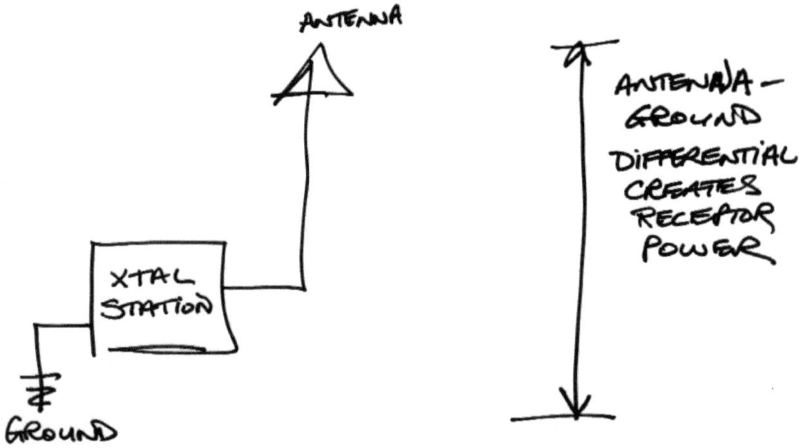

4. Now the Brass or Bronze Bell and non-conducting wood striker will go into their respective places. Make certain that the bell is not accidentally struck. It should only be sounded when operating the XTAL Altar after fumigation and candling (see below).

5. Place the sticks of incense into the incense storage case and then carefully put this assembled item into its proper place on the altar.

6. Place the incense burner onto its area and remove one stick

from the incense box or holder, carefully inserting the wooden end of the stick into the burner receptacle socket.

7. Now remove a beeswax tea-light from its package. Tea-light candles may be stored elsewhere than on the altar. Hold the tea-light candle over the incense so that the smoke from the incense rolls over the whole of the candle. You may have to rotate the candle to get the full effect.

8. Now carefully holding the tea-light level, use a wooden safety match to light the wick, and place the lighted candle into the Rock Salt Crystal Candle Holder.

9. Now fumigate the wooden bell ringer stick over the incense and while still holding it, pick up and fumigate the bronze or brass bell in the same manner, rotating the bell carefully over the incense to insure that all areas of the bell have been properly fumigated. Use of a mantram while fumigating is a very good practice.

10. Now fumigate the Rock Salt Candle Holder over the incense in the same manner, holding it carefully so as to not tip it in any direction and then replace it in the center of the altar. Do this with both candles on both end-altars, first the left, then the right.

11. Ring the bronze bell on the left altar three times, allowing the vibrations to almost die out each time, but make sure the sound is still hearable although faint before striking it again. Replace the bell in its spot on the altar.

12. Now it is time to remove your .5 to 1.0 KiloGram NWA Stony-Iron Chondrites from their shipping containers, and place them precisely on the altars from left to right positions. Make certain not to handle them overly much, as the most important element in these items is the Reichian effect caused by the combination of fusion crust and chondrite material within. Chondrite meteorites of this type contain inclusions from exploded stars as well as the earliest known matter in our solar system. You will not need to fumigate the chondrite meteorites, and it would damage them severely to do so.

13. Now you will fumigate the Crystal Tuning Spheres and place them on the altar.

14. It is time to place the Dunwoody on the central altar. Make certain that it has been assembled according to the instructions

given with the kit. Coil winding is especially important if your Interdimensional Radio is to function properly.

15. Place the Bucher XTAL QRM Rejector in its correct location, and connect it to the Dunwoody Set according to the directions given in the kit.

16. Now you will want to connect and place the Morgan XTAL Antenna Tuner on the central altar. Do not attempt to tune the antenna at this time.

17. The High-Impedance Amplifier/Speaker can now be placed according to its position on the central altar.

18. Having loaded the Hardwood “A” Battery Holder, place it into position.

19. Now load up the Hardwood “B” Battery Box and place it on the altar.

20. String the Diamond Magnet Wire High-Gain Antenna between its buffering nonconductors making certain not to operate the antenna during lightning storms. Proper grounding must be observed here, as explained in the antenna wiring kit.

21. Hook the antenna into the system. Do not attempt to operate the radio at this time.

22. Now fumigate and place the Ticklers and other Targets.

23. Connect the “A” and “B” battery systems to the unit.

24. Now it is time to find the optimum signal by using the cat's whisker.

25. Use the Bucher QRM Rejector to suppress any local radio broadcast stations.

26. Tune the antenna at this time, to get the strongest possible signal from distant stations.

27. Now tune the Dunwoody to find an area of static that has faint background sounds in it. You will know it when you find it. You can almost hear a voice, but it is not a human voice. Listen carefully and you will locate the exact frequency of your own Guardian Angel. When you hit the “sweet spot” you will recognize your Angel's voice without doubt.

28. Now hold the Resonating Bell above one of the Resonator Crystal Spheres. One of them will work better than the others, and this may vary from time to time. Strike the Singing Resonator Bell

three times lightly on its side, then five times, then three times again.

29. Begin rotating the ringer around the top edge of the Resonator Bell, stroking it firmly but gently to produce a low continuous singing tone from the bell. This vibration must be fairly steady to get the fullest effect. A break in the sound or buzzing from too much pressure on the bell will break the invocation somewhat or completely, depending on the time it takes for the singing bell to recover from the break.

30. When the Angel is operating and sending signal through your Interdimensional Radio, the bell will almost seem to sing by itself, requiring no effort whatever to keep it going. The ringer will barely contact the side of the bell and the motion will seem actually unnecessary. When the sound produces itself, the Angel is operating in that space.

31. You have now successfully contacted your Guardian Angel. You can ask questions, converse with the Angel and listen for responses coming from the other dimension.

32. You can use your Interdimensional Radio to contact and converse with those who have passed from human life into the Afterlife in exactly the same way.

33. Your Interdimensional Radio can also be used for Angelic Healing by charging your Ring of Power, Charm and Tickler Amulet, then bringing those items to the person in need of healing.

34. Angelic intervention is also useful as a method of dealing with emotional crises and moods of depression. In cases of clinical depression, medical advice should be sought. The Angelic Radio is not a curative nor is it a medical device of any kind. It is simply a means of contacting Angelic Guides and should be treated as such.

ZADKIEL'S PORTABLE WISH-LIST
POWER ALTAR

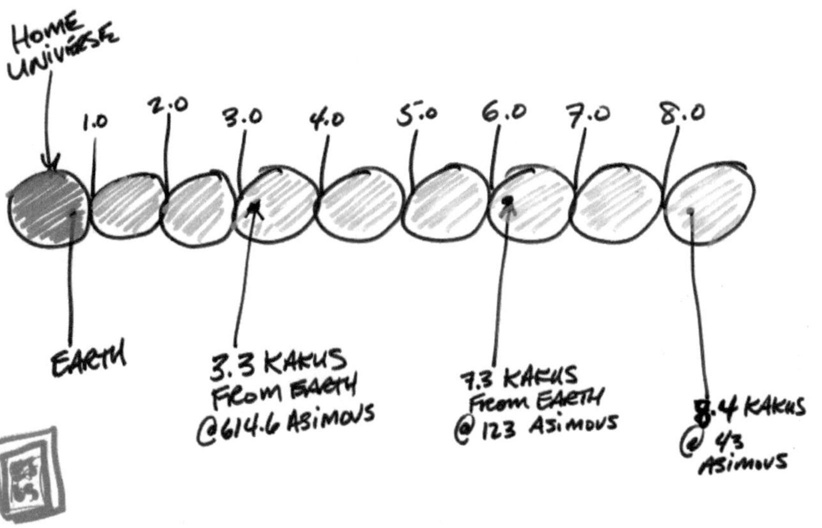
HOME
UNIVERSE
1.0
2.0
3.0
4.0
5.0
6.0
7.0
8.0
EARTH
3.3 KAKUS
FROM EARTH
@614.6 ASIMOVS
7.3 KAKUS
FROM EARTH
@ 123 ASIMOVS
8.4 KAKUS
@ 4/3
ASIMOVS

How to Operate Your Gorebagg Spirit Radio:

Connect the antenna, ground and amplifier/speaker or high-impedance headset such as the military type to the correct Fahnestock clips on the receiver. Start with the Antenna Circuit Clip on the coil tap at 70 turns and the Detector Circuit Clip Lead on the 100th turn tap.

Slowly adjust both air variable capacitors in unison through their ranges until you can hear several AM commercial stations blasting in. The twin tuning capacitors work best when they are tuned simultaneously and kept in the same relative positions. There are other settings as well, which you will find through experimentation in your radio zone.

The whole AM commercial broadcast band can be tuned in by moving the Antenna Circuit Clip Lead to various tap positions on the Coil, down to the 30-turn tap.

The Detector Circuit Clip Lead can be positioned on any tap higher or lower than the Antenna Circuit Tap Lead to find the best signal strength and tuning with the least amount of interference from other stations.

Short wave stations in the 41 and 49 meter bands will come in during nighttime conditions with the taps at the 5th, 10th, 15th and 20th coil taps. You may want to tune down even lower under some atmospheric conditions.

Signals will fade, drift in and out and boom or fail as the ionosphere charges and discharges during the night hours. It is normal to hear more than one station at a time, but you should be able to easily tune the conflicting station or stations out and hone in on your target Spirit Voice, which will sound very much like someone whispering right next to your ear.

Some Spirit Voices are stronger, but most sound very whispery and have a certain ethereal quality to them. You will soon recognize familiar Spirit Voices as you work with them on a nightly basis. Do not attempt to use your Spirit Radio during the daylight hours; it just won't work.

Short wave signals tend to skip and bounce off the ionosphere and may hop or skip like a flat stone thrown across a pond, coming from the far side of the planet, and even from space.

Ionospheric alternating charge and discharge is dependent on solar activity, sunspots and solar storms in particular, and on the time of day, seasonal changes, sunspot cycles and weather and atmospherics of all sorts, as well as broadcasts by amateurs and commercial stations.

Some nights you may not be able to detect a single radio signal, and other nights you will be able to hear stations all over the globe and beyond.

On quiet broadcasting nights, you have the best chance to hear the Spirit Voices because the interference is simply not there. In very crowded areas such as a tightly populated city, you might want to think about adding a QRM Suppressor and an Antenna Tuner, both of which are available from us.

AM Commercial Broadcast DXing, or long-distance reception and logging, is also possible, following the same general rules as you would for short wave transmissions. The primary problem with commercial AM stations at extreme distances is to pull in the weak signals while at the same time rejecting the stronger nearby broadcasts. Practicing with the Tuning Capacitors will give you the key to reduction of stronger stations in order to pull in the weaker ones.

In an extremely strong urban area with powerful transmissions all around you, it will be necessary to use your wave trap interference filter to eliminate the stronger commercial radio traffic and noise. Your QRM Suppressor will do the job, with practice, patience and a growing familiarity with the vagaries of radio wave transmissions in general.

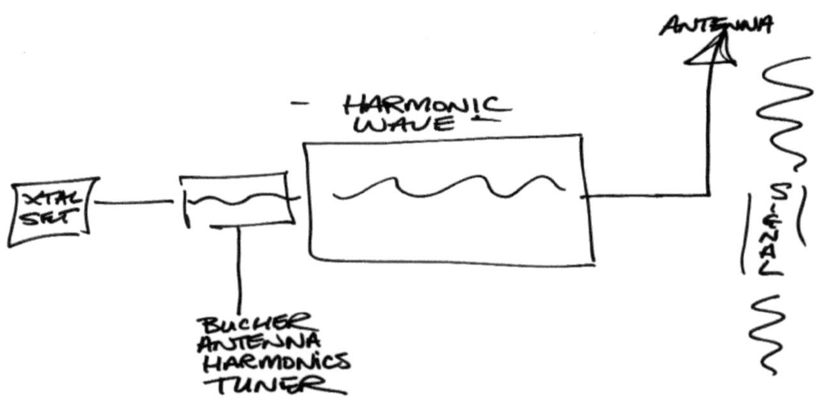

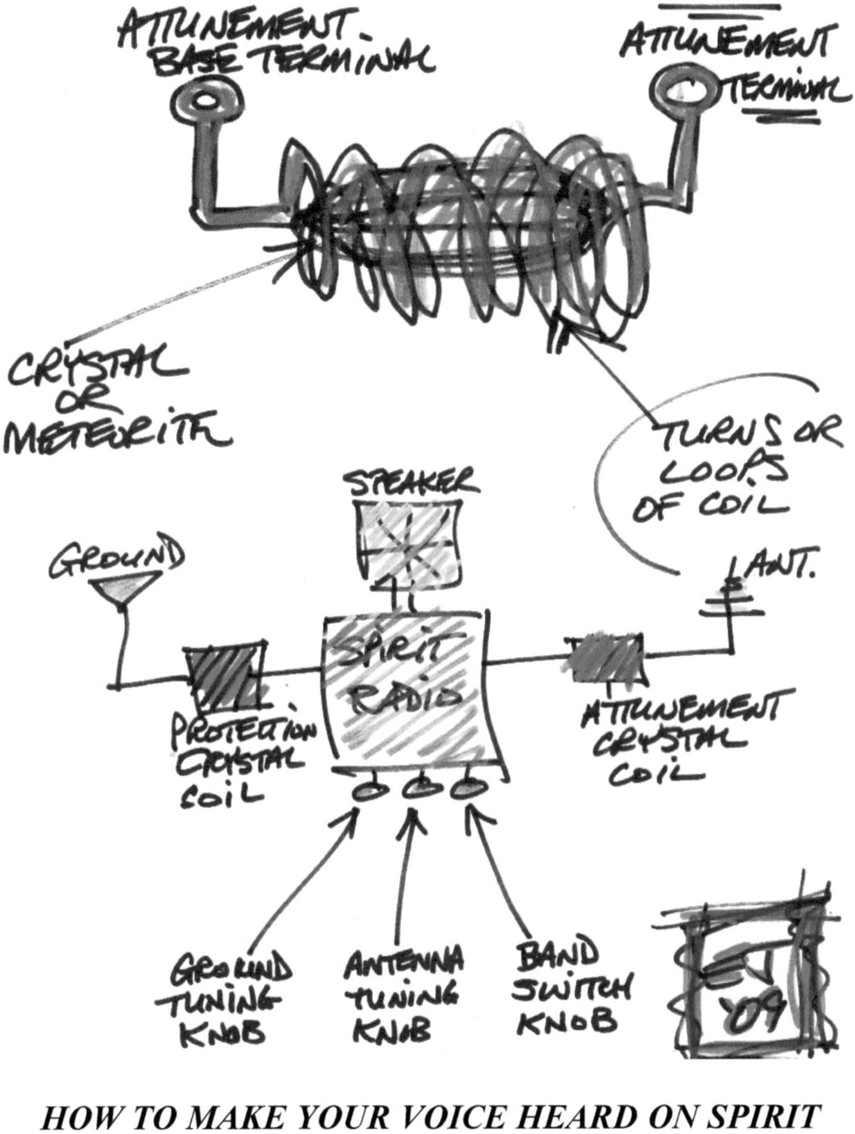

HOW TO MAKE YOUR VOICE HEARD ON SPIRIT RADIO FREQUENCIES

1. Use the Gorebagg Spirit Radio contact ritual, adhering strictly to the ritual.
2. Use only the correct instruments, incenses, oils, candles, crystals.
3. Use the correct power wands for Spirit Radio contacts.
4. Use the correct bell tones in the proper sequence.

5. Use the correct Gorebagg Spirit Radio Contact Mantra.

6. Make sure to bathe, cleanse and purify with the Gorebagg Spirit Radio Contact Crystal Bath Salts.

7. Make certain your GSR Operator's Contact Garments are fresh and clean using only GSR Crystals as a cleansing agent.

8. Wear the appropriate contact charm, contact ring, contact amulet, contact bracelets for the occasion and do not use them elsewhere.

9. Purify your intentions with high indifference as to the outcome of the contact.

10. Use only your Spirit Contact Journal and Spirit Contact Recorder to record the contact.

OBJECTIVE TESTS FOR SPIRIT CONTACT

1. On the Gorebagg Spirit Radio, a contact can hear your voice without the use of a radio transmitter.

2. Spirit Voice will respond directly and specifically to a GSR — Gorebagg Spirit Radio — operator.

Here are some warnings relative to contact with Angelic Guides:

Angels do what they do; they all have their own unique characteristics and obey laws which are totally inflexible. Therefore, make certain you really want what you ask for — you might just get it.

Angels are not the servants of human beings, nor do they have much interest or understanding in the plights and goings-on of biological life-forms. Nevertheless, their influence can be brought into direct interactive play in any given space and time and with their cooperation, amazing and actually miraculous things are possible.

Angels have both attitude and altitude. Angelic conversation and contact can be both enlightening and interesting, but can be dangerous if you want things for your own benefit, especially specific life-game outcomes.

Angelic invocation using this resonance radio actually works, as any psychic will agree; however, as in any contact with higher entities, extreme caution and fair play, respect for the nature of the contact and a good deal of common sense and a well-developed sense of honesty and humor will be your greatest allies in your quest to delve beyond the human dimension.

Nothing real “does itself”. You will get out of the XTAL Interdimensional Radio only what you put into it. GIGO, Garbage In, Garbage Out, really applies here.

Any experimental project beyond the ordinary perception is limited only by the imagination, resourcefulness and courage of the scientist/investigator.

Master the basics of Angelic Resonance Invocation and you will

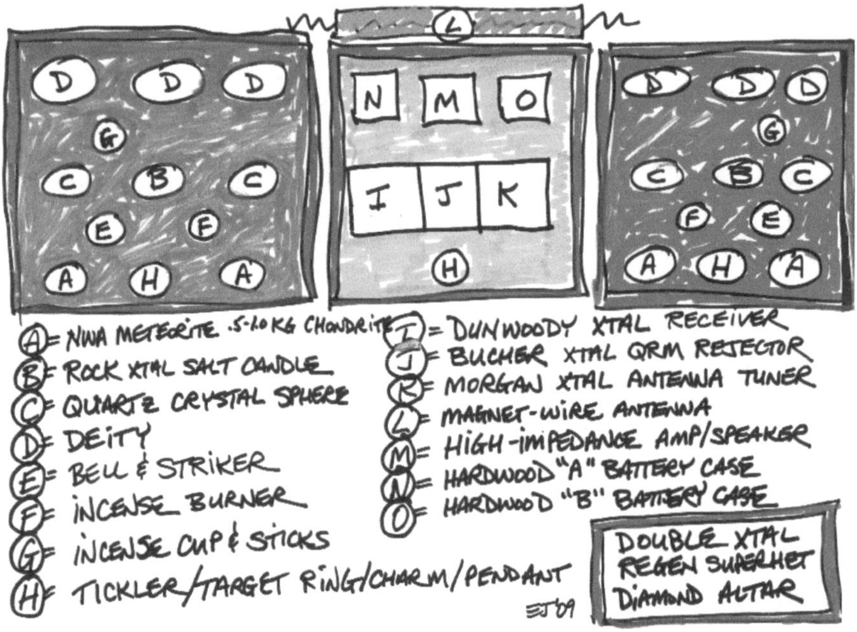

soon be thrilled by the potential inherent in interdimensional contact.

If you've ever wanted an objective recordable message detector when communicating with the Recently Deceased, Angelic Guides, Trans-Dimensional Entities and more, this is the altar device for you.

The Spirit Radio Altar is not easy to construct. It is a level of difficulty of 10 on a scale of 10, therefore I build them and ship them prefab, or you can construct it yourself with high level soldering skills and the ability to follow a circuit diagram to successful conclusion, combined with coil winding skills and lots of time on your hands — the project takes a couple of days to complete.

If you elect to send for the pre-fab assembled kit, your work is far easier, and takes just a few minutes to assemble on the altar table.

You merely arrange the various parts and follow the easy instruction manual on how to operate the Spirit Radio Altar in your own home.

The manual for the construction of the Spirit Radio Altar uses photos, diagrams and text to explain how it works, why it works and how you can learn to use it. It does not explain WHY you want a Spirit Radio Altar; that is the province of your own work path.

The technology to build the Spirit Radio has been lost for many decades and the right vintage parts are extremely hard to find.

The completed pre-fab Spirit Radio Altar Kit includes everything you need except the table itself: Claude and I will construct the electronics for you. Here's what you get:

1. Spirit Radio & Special Antenna System
2. 2 Spirit Radio Black Tourmaline Protection Crystals
3. 2 Spirit Radio Quartz Spheres
4. 2 Spirit Radio Chondrite Meteorites
5. Spirit Radio Handmade Wool Hand-Woven Altar Cover
6. Spirit Radio Candle Holder
7. Spirit Radio Incense Holder
8. Spirit Radio High-Impedance Amplifier/Speaker System
9. Carborundum Detector and Cage System
10. Spirit Radio Operator's Manual

Since it takes a couple of days to produce these, the orders are on a first-come, first served basis. We will construct the units as we are able with our very full workshop, game and book release schedule. We are currently working on six new books in the field of art, work ideas and game development.

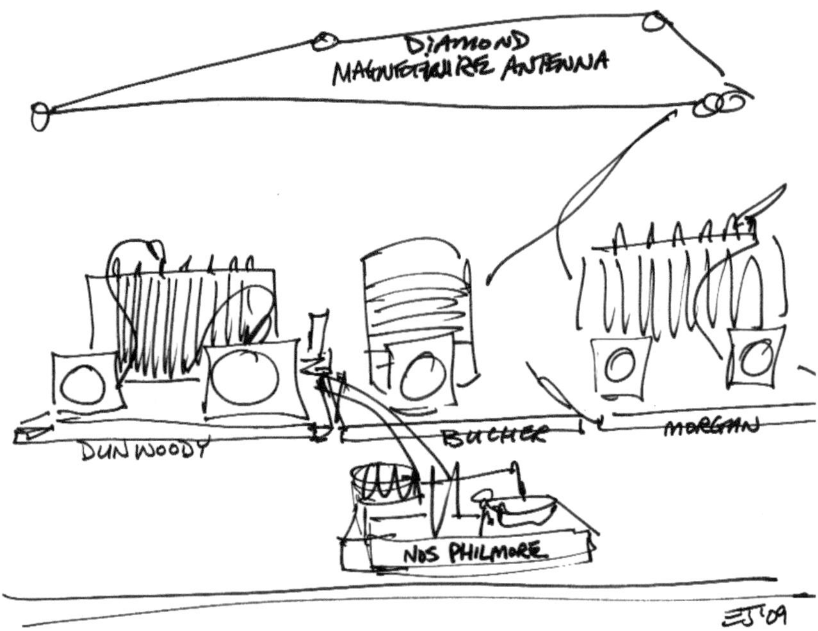
DIAMOND
MAGNETWIRE ANTENNA
DUNWOODY
BUCHER
MORGAN
NOS PHILMORE
EJ '09

ALTAR SETUP DOUBLE REGEN SUPERHET EJ '09

E.J. "Gorebagg" Gold gives workshop and demo on Spirit Radio, March, 2009

Spirit Radio workshop, March, 2009.

Chapter Four

INNER CIRCLE WORKSHOP ON SPIRIT RADIO

I wanted to talk about the Spirit Radio. It's sensitive to all radio frequencies and it will, through oscillation and resonance, tune into certain selected frequencies of the radio spectrum. However, along with this and its crystals it has a dual air variable capacitor system in the front and operates as a Beta-Blocker.

This currently has a Germanium diode, and I now have a detector cage and trap system, which is about a hundred years old. It's an unused antique Phillips Detector Cage. In it, I plan to put a crystal, but not a Germanium crystal. I'll be using a different kind of crystal. It's much more sensitive and much more difficult to mount, because the thing has to be mounted in a cold mount metal mounting. The metal has a boiling point, actually has a melting point that's below the boiling point of water. Lead's too hot and it gives the crystal a heat history that destroys its sensitivity. So it's a super sensitive crystal. It's probably going to be a Carborundum crystal that we'll end up with. So the crystal will be in a detector mounting with what's called a cat's whisker, which will find the most sensitive point on that crystal, hopefully.

I have some taps in this coil. The number of taps in this coil is an outrageous number, but I have ten taps, each of which determines the wavelength to which we'll be able to restrict the tuning.

Apart from the other crystals in the set, this is basically a crystal receiver. I have an Armstrong single tube receiver, but that is not going to do the same job as a crystal receiver. The crystal receiver receives all of its energy from the differential between the antenna and the ground. So the farther apart the antenna and ground are from each other, the greater the potential, the greater energy that will come to this receiver. It doesn't have any batteries. It, in fact, can't use batteries for the purpose. It doesn't generate power within itself.

For a crystal radio you need to have several things in addition to the radio itself. One of the things it turns out you need will be a gang of QRM suppressors, each one of which suppresses very powerful local radio stations. You want to tune those out so that you can "see" past them into the distance.

Essentially, that's what they're doing now to look at planets next to stars. They're trying to tune the star out so they can see the planet or planetary system around it. They've found many, many, many, many planetary systems now around stars in our local space/time continuum well outside the Solar System, but as they search more and more for them, they discover more and more the need to tune out the suns, or the stars, that are at the center of those so they can see the planets because the star obliterates the light and also other emissions from the star. That is to say, x-ray emissions, radar emissions, different emissions in various areas of the spectrum, electromagnetic spectrum. It's throwing out all kinds of things. So it's blocking or blotting out the smaller, weaker signals.

So in the same way, the Spirit Radio has to block the stronger signals, block out all the local radio stations. So five or six QRM rejectors in series are going to be necessary. Then over here will be the antenna tuner, which is actually a way of receiving resonance within the antenna system, so that the antenna is resonating at the highest possible energy level, energics, or energetics level. "Erg" is a unit of energy. So an energetics booster is achieving the highest possible resonance with the antenna and the receiver, trying to get the best energy-balance between the two.

One of the examples that's used quite commonly in talking about resonance is the bridge, where you take a span bridge and you either imagine soldiers walking across it or imagine hitting it with a hammer. Now, the relationship between frequency and wavelength is a very solid one. If you hit the bridge just one time, you push the bridge down slightly, but if you push it down, the bridge comes back. So that is a single wave. The bridge goes down and then it comes back. As you hit the bridge, you'll hit different frequencies.

Let's define frequency as the number of events that occur in a given amount of time. How often it happens. Frequency — how frequent is it? Frequency is the number of times that something happens within a given span of time. Wavelength is the length of the wave measured by meters, feet or any other linear scale.

In fact, you can run this experiment with a rope. You use a skip rope and you can tie it to the handle of a door or, if you have a compatriot you can work with, they hold the other end, and you just make the rope go up and down. Now at one frequency, or the number of times per unit, say per minute — you go up and down and up and down — per second works well with a rope, I think . . . so if you watch from the side, you'll see the rope making arcs that go up and down, up and down. Now if you go faster, that arc gets split in the middle and then you have two arcs that are going up and down, one opposite the other.

It happens twice as fast and the wavelength is half the size. Now if you increase the frequency again, you can either obtain 3 waves or 4 waves or 8 or 16 or 32. If you go to a guitar string . . . this is the most amazing thing to me. Pythagoras, in several hundred B.C., noticed that if you twang a guitar string, the equivalent thereof, a zither string, a harp string, an instrumental string, if you twang it, it will vibrate at such-and-such a number of vibrations. Exactly how he measured that, I'm not sure.

We know ways of measuring with crude instruments — attach a pencil to it and drag a graph page across, and so on, but somehow he made this intuitive leap that if you put a bridge of some kind somewhere in that string, it will divide the string's vibrations up in very specific ways, and it always does that division in exactly the same way, no matter whether the string is fat or thin, long or short. It will divide it up in the same variables, the same ratios. Not the same numbers, not the same frequencies, not the same wavelength, but the same ratios. So if you cut the string, if you bridge at the half-way point, what happens when you twang it?

You'll double the frequency and you'll get equal frequencies on both sides of the bridge. Now you put it to two-thirds, one-third, and the wavelengths will vary according to that exactly the same and the same ratio 2/3, 1/3. And so he noticed that you could tune an instrument such as a guitar — at that time it would have been probably a Kamanche which you can hear on youtube.com by Nune Yesayan and composed by Sayat Nova. You could tune by moving the bridge as well as by turning the tuning peg, which is what a capo does on a guitar. You actually retune the güitar by moving the capo up and down your frets. That is the basis, actually, for Pythagorean theory. And it's one of the ways in which Pythagoras was able to derive such things as the theory of the isosceles triangle.

The nature of the frequency is such that if the length of your string is fixed or the length of your antenna is fixed, the coil can be variable, because we put taps every so often with clips, and mount a shorting switch so you don't have to keep changing the clip — what you do is you simply wire each of these taps to the shorting switch — then you turn the shorting switch in order to select the various radio bands that you're going to want to listen to.

For instance, if you're listening to interdimensional radio traffic it occurs in a different band than traffic from passing spirits, what might be called dead spirits. Also, you're going to pick up spirit-based Theta and Alpha communication, which is communication from very deep levels of the mind. Those levels will be on totally different bands, so you want to actually switch bands. Band switching means you'll make this coil shorter or longer using the taps, and the shorter or longer the coil locates the different places you'll be tuning into along the electromagnetic spectrum . . . .

We should talk about the spectrum in terms of size of wave sizes from extra long waves all the way down to short wave. Short wave, by the way, means the wave is shorter. Long wave means the wave is longer and short wave means it's shorter — longer wave, shorter wave — extra short wave, ultra short wave, ultra high frequency.

The faster you go, the shorter the wave has to be . . . the shorter it'll become.

Light speed is not a frequency, it's a speed. A photon has whatever wavelength it has. A photon can have a wavelength of, in ELF — extra low frequency — the wavelength can be measured in hundreds of miles. Photons tend to stretch out the further they travel.

Is a photon energy or mass? Yes, both! It's transferable. It's either measured as energy or as mass. It's either particle or a wave, depending on how you measure it. Wavicle is what you call it. What is a wave as opposed to a particle? It goes by behaviors.

It's very similar then, in some ways, to electrons which can be divided by behavior — upness, downness, strangeness and charm.

If you create an experiment that measures the ability of a photon to act like a little miniature beebee, then what you see is a photon acting as a miniature BB.

I've got an interesting question here from Piph: "Disembodied spirits that hang out around this atmosphere are essentially a wave or radio wave or wave and transmitter?"

Hang around this atmosphere is not necessarily correct. Space and time are an illusion, lunchtime doubly so. Space is curved and creates the illusion of gravity by being curved. What it's bent by, leave that alone for the moment; how it's bent, leave that alone for the moment; but space *is curved.* It creates the illusion of gravity, at a gravity well. Location has nothing to do with transmission. And radio transmission especially so. Radio waves travel at the speed of light. The way that a spirit communicates is usually they don't. Generally a disembodied spirit is lost, non-locational and non-identified, and therefore, doesn't have any wish, desire, interest in communicating, or any knowledge of how to communicate or with whom to communicate. Once in a while a disembodied spirit will wake up, as it were, in nowhere, the 11th dimension. We could call it a thread or string.

One way a disembodied spirit decides to communicate or responds to communication, could be in a séance where someone is calling to them, "There are questions that only you can answer. Come in please spirit of Zonk O'Brien." And the spirit of Zonk O'Brien answers back, "Huh? What? W-w-where am I?" Well, that's a question in itself, isn't it? The answer or the comment or the communication from the disembodied spirit occurs because in that state, in the between-lives state as well as from another dimensional state, it is possible to influence — I don't know how to put this — to influence voltage, I guess, or influence electrons, to kind of move them a little bit, to bend space and time, just slightly, enough so that a phase form, a phase factor can be introduced into the wave form, and that's what we're talking about. We'll get to the phase factor later.

But what happens is the disembodied spirit actually begins to meddle with the phase factor of a wave or a particle or a wavicle, which is to say, you're only describing the behavior of the thing, not the thing itself. And the same way that a quark only describes the behavior of electron, it is not in and of itself a thing.

It has to do with what you look for. Most of you who are listening to this broadcast are *somewhere.*

There are disembodied spirits also listening to this broadcast. That's one of the reasons we do these broadcasts.

So, wherever you happen to be, there are going to be blue, there are going to be red, there are going to be pink and orange, there are going to be various colors. And black and gray. Now if you imagine putting on a pair of glasses that only allow you to see blackness, then whiteness, greenness — anything other than blackness, you can't see.

How about a pair of polarized glasses that only shows you the x-ray range and let's presume you can see in the x-ray range, or the radar range?

The difference between particle and wave is that it turns out that when you wear glasses that only see waveness and you look at a photon or an electron, then you will see the photon or electron as waveness. However, if you wear glasses that only see particleness, then you see these photons and electrons as particles. Now one way to look at something as a particle is to ask a question such as: "If I bounce this into something, how will it be ricocheted?" As soon as you ask the question about ricochet, or bouncing off, you're talking particle, you're talking little teeny billiard balls or beebees. But if your question is something like, "If I put this through a defraction grading that bends waves, how far will they bend?"

Or a distribution block.

It's a double slit which measures based on frequency and based on the wave nature. You're asking a wave question. So if you ask a wave question, you're going to get a wave answer. If you ask a particle question, you're going to get a particle answer. Now if you ask a wave question to a potato, you're going to get no answer. If you answer a particle question to a potato, you will get an answer. Potatoes are basically particles. They're lumps.

It's not just a radio, it's an entire station. It's what is called by Ham radio operators a "Rig."

The early Marconi sets, are "home brew." People made their own radios at that time, a hundred years ago, when radio was just starting out. Radio was then where computers are now. Now!

In the future, if it's possible, people will be going back to today's computers in order to accomplish things that the future computers will not be able to do.

And that is exactly what happens in the case of modern radios. Modern radios cannot operate on the spiritual bandwidths in the way that a crystal radio can.

The stations that you are able to receive even on the AM — amplitude modulation — band are digitized, quantized.

Here's a digital radio. Press the button, it goes to the next strongest station. That's what the digital radio will deliver to you. It won't deliver what's right next door to it, and it certainly won't deliver what's behind it at a distance.

You've got a — let's call it a star — and it's very close. If you could move that star aside or bend light in such a way that you could see behind it, you would notice that there is a little speck behind that star. Now that little speck is only one trillionth of the diameter of the star and it's one trillionth the amount of light of that star, the illumination of that star, the luminescence of that star. So that number of lumens relative to the lumens of the stars can be very, very slight, very small. But let's say you had some way of moving the star aside and looking behind it. And if you did you could see behind it a little tiny speck. It's very dim. What you might not realize is that that's not a star, it's an entire galaxy. It's so far away that the light is barely visible. At least the light in the visible light spectrum is barely visible.

So when they turned on the Hubble telescope, that's the first thing they notice is that there are a lot more galaxies out there than they ever thought there were. Not just stars, but many of the things they assumed were stars were actually galaxies and super galaxies. When you move aside the local radio stations, behind those stations are very distant stations. When you move those aside, or to and around them, bend light around those, or look right next to them, you will find micro assignments of frequency in which disembodied spirits or extradimensional entities are actually warping what we call radio waves. They're warping the electromagnetic spectrum in a particular frequency. It could be 40 meters, it could be 2 meters, 6 meters — that's the size of the wave, if you know how to look and where to look, and how to move aside the big guys, the really big guys, like that hammering 50,000 watt station that's four blocks down the street.

If you could move all those aside, and you can, you'd see extra distant, very distant and ultra distant weaker signals that are at the exact same frequency as the bangers. This device is for that purpose. That's a QRM rejector, which rejects the very strong signal strength by tuning it out, literally by canceling it out with itself. It's called wave cancellation, which happened to be my specialty — wave cancellation and beat frequency oscillation. That was my specialty. Claude had his own specialty, which was looking at the phenomena that occur when a packet of low frequency long wave items combine together in order to create a high frequency result, with sub-packets and leftovers.

That was what he was looking at in his research, and he was applying this to proteins using the heat bath that would create long wave low frequency vibrations that could then be lumped together and made into an electronic or higher frequency transition, which would be equivalent to taking four radio stations, adding them up and making visible light.

What he was doing was using some tractor frequencies and the radio waves which he would lump together a number of frequencies and let them crash together in much the same way that you might hit notes on a piano and let them crash together to make a composite wave. The composite wave had leftover waveness. The leftover waveness was what Claude was interested in. The subtractive frequencies leftover.

My specialty was similar to Claude's. What that led me to was the discovery of the Spirit Radio in which you're looking at subtractive frequencies rather than the strong powerful obvious frequencies. You're looking behind the ordinary transmissions. You're trying to hear something which is normally buried by the bangers, the heavy bangers that are blasting out their radio signals as powerfully as the government will allow them to.

So between these two, we came up with something, which is kind of interesting that actually John Lilly worked with as well. John and Toni used transducers to communicate with dolphins underwater, because the transducer translates an audio signal into a waveform which is transmittable in water. So one of the things that we noticed is that there are corresponding octaves in the light range and harmonics of these in the sound range.

We saw that later on with radar and radio, and so forth, and higher and higher and lower and lower frequencies. But actually what happens is, in a sense, you can get a D minor chord or an A 7th chord in light. You can strike that chord in light frequencies. On a light range of electromagnetic spectrum or in the x-ray section or portion of the spectrum. You can actually strike a chord like that. There are correspondings to the audio, or I should say the audible, the human audible sector of the EMS.

So there were correspondings all along the electromagnetic spectrum. There are repeating octaves which go up and down the electromagnetic spectrum. So if you hit one chord, let's say, in audio frequency and you happen to get the harmonic, the harmonic frequencies will strike correspondingly in the light frequency range, and also in the x-ray, and in various places along the electromagnetic spectrum. So you can actually create harmonics, wave harmonics, that occur.

So spirit communication is in very, very high frequency and very ultra frequency. You can't read those, so what you're trying to read is the wave harmonics in the lower frequencies. So we tune in around the broadcast band, AM broadcast band, and that's where we can pick up the harmonics quite well. It's got a little more drift than it would have in the higher frequencies. It's a little harder to understand, but you can still pick them up and you can still actually read them quite well.

We talked about a skip rope and waggling your arm up and down so that you can see waves, literally waves — the rope goes up, the rope goes down.

It's an experiential thing that you can understand by looking at it, by actually experiencing it. One of the problems with non-Einsteinian and non-Newtonian physics is that you have no experiential proofs, no experiential data that you can rest upon. So it's hard to get hold of quantum mechanics, for instance.

It's also true that if someone sitting out there has never seen a skip rope, has never played with a skip rope, they're missing the experiential data for that as well. And unfortunately there is less and less experiential data being picked up by individuals.

Something that came up a little earlier that we haven't touched on and I don't think I want to touch on it too much here, but the concept of your personal suffering, or your personal misery, what I said was that you will never give up your personal misery, your personal suffering, which is actually a body of suffering, a body of misery that you have created around yourself. Before you were five years old you made this thing. It's composed of your experiential data up to the time you were about five years old where you learn how to handle life as best you can, how to handle organic life the best way you know how, the best way you learn how, which includes a certain amount of suffering, a certain amount of misery, a certain amount of unhappiness, a certain amount of gloom and doom. And this is absolutely the last thing you're going to give up because it actually creates your comfort zone, believe it or not, because it's understandable, it's predictable, you know the crap that's going to happen to you. You know exactly what bad things are going to happen to you. You keep doing those bad things to create those bad things because it's predictable and it's knowable and it's your comfort zone. You're very happy in your misery. You're wallowing in a pig sty that that pig sty is of your own making and you know what's going to happen within it.

That actually creates the tendency toward automatic rebirth because you're going to recreate that same misery from the between -lives state. You're going to want to create that same misery, so you're going to go find the same crap life and go through the same crap things because you're familiar with it. It is your comfort zone. So your personal misery is the last thing in the world you'd ever want to give up. In fact, even if you wanted to give it up, because you're so miserable in it, you wouldn't know how to go about that. So part of our training is to give you the skills to eliminate, to cancel out QRM rejector. To cancel out your personal misery, to cancel out your brand of comfort zone. That means that you have to be willing to live outside the comfort zone, which is going to create a new kind of misery. But that misery is the misery of not knowing what's going to happen. It's like forever waiting for a hospital or a medical report to come back from the labs . . . forever waiting, you'll never know what's really wrong with you. So it's waiting for a lab report forever, many lab reports forever.

So it's undetermined, undetermined, indeterminate, indeterminate . . . and it goes on forever. It's never going to stop like that. That's the way it will always be from that point on, from the point where you lose your personal misery, your personal suffering, you will never really know what's going to happen to you, or what's going to happen next. You'll never know. And that is produced by a QRM suppressor. In fact, everything in this radio system has a corresponding in work, so that detection of signal and reception of signal, and selectivity, and so forth, all of these things actually have meaning and significance in relation to your work.

So building one of these things and working with one of these things will actually teach you a great deal about how to accomplish that within yourself. I don't mean actually winding the coils, but that's what we'll do for you. But actually by assembling the radio and working to contact and listen to disembodied spirits and extradimensional communication will actually help you to understand how to transcend your personal misery.

So as you wiggle this rope up and down, let's say you have a wavelength of one half the rope length. That means that if you look at the rope from the side you'll see half the rope going up and then it goes back down to the line in between you and the friend or doorknob.

That's the zero point.

Then it goes down and back up. So zero down zero up zero.

On an oscilloscope that would be the minus and the plus.

Now if you take this wave and shift it one-half wavelength, the wavelength is all the way through the entire movement. So if you shift it half a cycle and add it together, that means you're going to have one wave that's going up, but then the duplicate wave, which is half a cycle out, is going down at the same time this one is going up. And then when this wave is going down, the half cycle shift is going up. So if you add these together, the up here and the minus here, the down here, they add together and give you a flat line.

Which is called wave cancellation. But it never occurs exactly — under some lab conditions it can occur exactly, but normally in nature it doesn't occur exactly — that the waves cancel each other flat.

In fact, they can't cancel each other flat.

It's theoretically possible.

It's mathematically impossible.

You can theorize that it's possible, but you can't make it happen, because it's impossible to produce a wave that does not have harmonics and subtonics.

As soon as you have a rope with two ends you have what's called end effects. And the end effects automatically generate harmonics that are high and low. That's through jiggle and wiggle and wave.

So you have the jiggle effect, which is that no matter how carefully you manipulate the rope, there will always be some vibrations that enter into it that produce inner harmonics, that is to say, produce subwaves within the primary waves. Those subwaves also produce subtonics and overharmonics, antiharmonics, and so forth. So that was actually my specialty to produce not the perfect wave cancellation, but to see what happens after those chordal crashes occur and you combine and crash and oscillate two or more waves together to create as close to beat frequency as you can. Beat frequency oscillation means the resultant wave is occurring slightly as opposed to very largely occurring. So those little tiny things are actually ghost waves; you could call them that. And those waves are what disembodied spirits can manipulate to make minor changes in voltage, minor changes in amperage, minor changes in amplitude, minor changes in frequency, they can modulate those subtonics, those little subtonic waves. Not the big ones, but they can modulate the small waves. The small waves are only detectable through such devices as the crystal radio.

They're not actually, in a sense, changing what is, they're changing what comes about as you subtract what is from itself. The wave is the wave. It's an actual thing that's banging through the atmosphere. It is immutable.

However, in your device, you're taking that wave and you're splitting in half almost exactly, but not exactly in half. And then you're offsetting it — oscillating it — and as you do that, it creates a difference — resonance. Well, that difference is not what is, that's what's being created in the moment, which means that even though this wave may have been floating around for who knows how long, is not being changed in itself. It's only right there, right then, as you are splitting and canceling that the way itself crashes against itself, hence, you have a very dynamic instantaneous and mutable — changeable — phenomenon occurring.

Something can act upon it. This is where Schrodinger's cat comes in and this is where we get into quantum mechanics, where the observation of the thing — this is another point — is that observing the thing changes the thing that's being observed. Because that's exactly how a psychic medium or a spirit medium works is they tune into what's left after they eliminate the primary signals. The whole point of the spirit medium — and I've studied this quite a bit, and I have conducted thousands of séances, not garbage séances, real séance — what you work to do for the first fifteen minutes to half an hour is you're trying to eliminate the local stations. You're trying to eliminate all the signals that are present normally in the space. Some of them are coming from the participants, some are coming from outside, some of them are coming from quite far, but all of those things are what are called "dominant waves," and you're trying to get those dominant waves out of there.

I worked with Hershel Toomim and John Lilly. Toomim used to hook me up to his biofeedback devices and have me generate Theta and Alpha, which I could do instantly for him — he didn't have to wait for it — so that he could determine what I was doing. What I was doing was I was suppressing Beta, which is the normal think think think think think think think. There is Beta one and Beta two and all of those associative mind patterns were gone. I was simply doing it by way of cancellation. I was canceling out the thinkingness and landing instead, or revealing instead the intuition, which is called Theta and Alpha, which is actually where ESP occurs, where your telepathy occurs, and so forth. So if you want to think of it that way, what you're doing is you're tuning into telepathic communication by eliminating the Beta and focusing on Theta and Alpha.

There is a tentative hypothesis that I have about how some of this works in that we are to a definite extent in a shared universe.

You've probably noticed that in the universe that you're in there's a lot of individuals that have very strong opinions about what absolutely, categorically cannot happen. If you're looking for a phenomenon that you can codify, carve into cement and use to show them that they are wrong, it's not going to pass muster.

Well, you can do it, but they won't accept it. Because even if you could codify it to human standards, it doesn't match their particular brand of suffering. It doesn't match their brand of belief system.

It's like they hold in place a suppressive field that prevents anything that they would truly understand and accept from occurring, which means that if you're going to hear a spirit on the radio, it's going to happen in such a way that they will not be convinced, which means it needs to happen as a clashing of frequencies or subtraction of frequencies or an interpretation where you bring your observation in a Schrodinger fashion or a Heisenbergian fashion, which I think is also very close to what's going on in the mix.

Typically, in science, if the facts don't fit the theory, you simply ignore the facts or throw them away. And that is the case, it's true. And that's exactly what is done. The last people in the world who actually accept a new theory are the scientists. They reject and resist as long as they can, just as they made fun of Einstein when he came up with his unified field theory. They immediately rejected it. Then they rejected it in 1913, 1915, 1918, 1921, they finally gave him a Nobel prize, not for that at all.

The fact is that scientists don't accept anything truly new. What happens is if you come up with something that is truly new, and you manage to prove it, they will give lip service — the living scientists — to that, and they will "accept it," however, they won't. If you look at why quantum mechanics had the growth and the evolutionary process that it had, it's because it took a certain amount of time for the old guard to die out. You started with Newtonian mechanics which were little billiard balls bouncing around.

Then, somebody came along and they said, "No, no, that's not quite what it's like. It's like this other thing." Well, a very small percentage of the scientific community accepted the other thing as it was. When it was eventually proven that engineering worked with this new theory, they sort of accepted it, but they didn't actually. It was only 20, 30, 40 years later when the "establishment" scientists dropped dead that the next evolutionary step in the theory could occur. Hence, you have these jumps where you have a new theory, and then they grind away at it. Then there is suddenly a new theory about thirty years later.

The theoretical physicists are like the receptacrystal. They're receiving all kinds of data, all kinds of information, all kinds of things. Within the circuit are resistors. Their resistors are the scientists, the educators, the people who want to keep things as they are. They have a vested interest in doing so. And more power to them. It's fine because it tests the theory; make sure that the theory is sound and solid, but the public doesn't see it or hear about it for at least 100 years. A hundred years later the public is finally starting to get hip to what happened a hundred years ago in physics.

There will be relativistic devices in use for a hundred years before anyone sees them as anything other than Newtonian. In fact, virtually everyone on the planet, when they think about an atomic bomb, sees nothing about $e=mc^2$. What they see is big dynamite. In fact, nuclear weapons are measured in tons and megatons of dynamite as opposed to energy and cause of energy levels, and most people cannot pronounce the word "nuclear" — they say "nucular".

What you have here in the Spirit Radio is essentially an etheric radio, a radio that tunes into ether signals. I will be laughed out of any scientific journal or meeting place by proposing that what you're hearing is actually extradimensional or extraorganic communication. I think we all have seen these, a number of programs recently over the last five or six years, there have been a tremendous number of programs about UFOs. Looking at those things and watching those programs would make me entirely unconvinced. I would be left completely unconvinced because there are cogent reasons why none of this can be true. And if it is true, it may be so that this happened. It may be so that this saucer-like craft did hover over such-and-such a place, but that doesn't mean it's extradimensional or extraterrestrial, or whatever it might be. It's some sort of obvious phenomenon of some kind that's understandable, if only we knew what it was. All well and good.

I would believe this except that I have had three personal experiences with extraterrestrial UFOs. One of those experiences was when I was in the U.S. Army and I was among hundreds of people who experienced and reported the event. The event was over a period of 20 hours. It wasn't something that happened in a couple of minutes and it was gone. There were photographs taken, there were films taken. None of those films or photographs have ever surfaced. There were official records made. None of those official records have ever surfaced. That's just the way it is. That's the way that the government operates. If it's anything out of the ordinary, it's going to be labeled as Top Secret.

The basic laws of physics are actually still labeled and classified as Top Secret in the U.S., not just military, but in the Department of Defense. Technically, you can't say $e=mc^2$. It's actually still considered Top Secret. There are so many top secrets and so many confidentials and top secrets and ultra secrets and so forth, in the government, that it would take millions of years and millions of workers to declassify the things which should obviously be declassified. There are things that are classified from the Civil War. I'm not joking. From the Civil War, the United States against the Confederate states Civil War. These things are still classified. They have no business being classified. But they still are classified because there is no one there who can decide whether to declassify it or not. And if they could, they couldn't spend the time and effort and energy and money to declassify these things. So there are things like Newton's third law of motion is classified Top Secret in the U.S. Government. And so, that is the kind of turgidity through which you're trying to wade. And so, I don't even bother, I don't try to convince the scientific community of anything because the scientific community is dedicated to stopping anyone from investigating anything new!

So, the way to get around that is to become really old school in terms of being a scientist. Which is what Dr. Hook suggested.

Because in the fifties, at least, they were teaching the scientific method. The scientific method starts with observation. It doesn't start with knowing the current system of theories. It starts with observation, then you propose a theory to cover what you're noticing. In that moment you might refer to other theories, but it starts with observation. Simple observation, it's called.

There are many folks who need to be convinced mentally or theoretically before they do the process of making the simple observation. And the proposal is, or suggestion is, that you'll do better if you make the simple observation and then you decide for yourself where to go from there.

You may not even be able to theorize. For instance, I noticed that when I perform the wave cancellation and beat frequency oscillation procedures with this particular equipment, not with digital equipment because digital equipment, as I said, doesn't work, tube equipment doesn't work, integrated circuitry doesn't work, transistorized equipment doesn't work. Only the crystal works at this time. I have theories, I have ideas about why this works. I have no idea how true or how correct or how realistic my theories are. I don't really know, but I can tell you from the point of view of pragmatically speaking.

When I perform this action, and then this action, and then that action, and then I put it through this thing, I get such-and-such a thing. And I have ideas, as I said, of why that happens, but I have no idea whether my theories can be ever proven or not, and I don't know whether they're valid or not, but actually from an engineering point of view, it doesn't matter whether my theories are correct or not. My practice is correct. In other words, what I do to produce the results is correct. And I get those results, and the results themselves are provable and demonstrable. What I cannot do is prove my theories about why these things happen.

One of the values of a theory, and actually one of the true measures of a theory's accuracy is, can it happen to better what you're doing? So if you make up a theory, or stumble across, or learn a theory about how the QRM operates, does that theory allow you to make this work better, or tune it in better?

I have a recording which was recorded under laboratory conditions with lots of witnesses. I asked some questions of some disembodied spirit and the disembodied spirit seemed to be lost in some way. I asked the disembodied spirit what the last thing is that it can remember and it says that I was on an airplane. Do you remember the flight number? Yes, yes, the flight was Flight 91. And so on and so forth. I don't know what Flight 91, but it may be significant to some people. I don't know of a Flight 91 that went down. But there must have been such a thing because the flight numbers, of all the flights that have gone down, a 91 must have gone down at some point or another.

So Flight 91, and there is this disembodied spirit, so somebody said to me, you know that could be a prankster who was playing with you by transmitting this back to you and pretending to be a disembodied spirit. I said yes, that's possible, but I'm talking into the atmosphere. I don't have a transmitter. So how would they hear me, and then respond to me, presumably from a distance, with their transmitter, and so forth? So that is not absolute proof. There is no such thing as proof absolute. Maybe there was a coincidence. Maybe I was talking and they were talking and then we happen to coincide, coincidentally talking about this one and the same thing at the same time. And the answers and responses coincidentally . . . all right, let's take coincidence, but pure coincidence on that level to me is just as interesting as spirit communication. Let's say it's pure coincidence. That's terrific too, if you want to communicate using coincidence as a tool.

This is from Grant, "Higher harmonics and certain corresponding lower frequencies is where the spirit communication is?"

No, it's the lower harmonics of higher frequencies, not the higher harmonics and lower frequencies. We are listening to lower harmonics of higher frequencies, and we're actually not listening to the frequencies of the harmonics themselves; we're listening to crash harmonics, which is to say, we're listening to the results of the diminished wave. The diminished wave is the wave which is left over after the waves have crashed and been subtracted, and so on.

Tamara asked, "Did anyone understand why or how the dead interfered with the phase factor?"

We haven't discussed the phase factor at this time. We should probably do that next time.

That's why you're not getting that point is we haven't covered it yet.

Tito says, "Does transcending personal misery mean being in the world but not of it?"

No. We'll have to talk more about transcending personal miseries later on. That'll be something which we will discuss. In fact I'll have a whole workshop on phase factor elimination of your personal misery and how to go about that. But we can't really do that until they have some experience, actual hands-on.

That's why I recommend very strongly that you investigate, experiment with, the Spirit Radio because that will give you hands-on understanding of the process which takes place which you can then interiorize. But you need to have it out there in your hands first before you can interiorize it inside yourself. It's just the way things work with you, with one, which is to have a hands-on experience of something, actually move things around, to push the clay around a little bit, to sculpt something. When you're sculpting in clay, you can then understand what I mean by mass. When you're merely talking about sculpting and clay, you can't really get it. It's all just nebulous ideas, nebulous concepts.

Your first big mistake is asking the kind of question, "Why should I investigate the Spirit Radio?" Well, that's your first mistake is asking that kind of question. You're asking me for exercises and I'm giving you exercises, so don't complain if you can't get any results. These are the exercises I'm giving you. If you don't do the exercises, don't blame me when you get no result, when you're still sitting there wallowing in your personal misery at the end of everything, at the end of life . . . and you're going to die. That's not the time to say, "Oh, maybe I should have tried the Beacon!"

Maybe you should have tried the Beacon!!!

Chapter Five

ONE TUBE VINTAGE SPIRIT RADIO

Upside:

1. Stronger signals
2. No coil taps
3. More stations
4. Higher sensitivity
5. Less chance for signal drift
6. Well-designed circuit with capacitors, resistors & tube
7. Less costly than cat's whisker crystal set
8. Less need for QRM/Tank Antenna Tuner Systems
9. Better audio output
10. Easier to tune DX stations and hear faint voices, i.e., spirit communication.

Downside:

It doesn’t work as a Beta-Blocker.

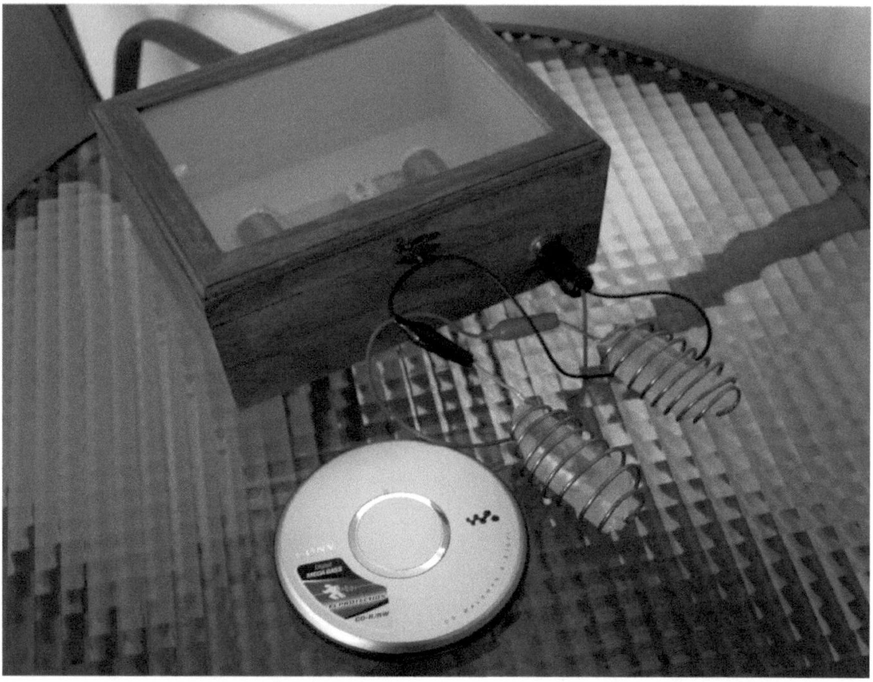

Parallel World Induction Device with crystal grips and CD player.

Testing the Parallel World Induction Device in the upstairs studio.

Chapter Six

PARALLEL UNIVERSE INDUCTION DEVICE

Based upon my work with the Spirit Radio, which was intended as a device for communication between dimensions or Parallel Worlds, I have stumbled upon a mechanical device that works well for what is commonly but incorrectly called "Astral Travel," or "Waking Dreaming."

Astral Travel is what happens to you when you fall asleep, meaning the Beta 20 cycles per second brainwave functions of the associative and rational mind diminish greatly and one floats downward through Alpha 9-14 cycles per second range into the Delta regions at about 4 cycles per second, then spiral upward again into the Theta Range, at about 8 cycles per second, a perfect wavelength for fully awake and aware interdimensional or Trans-Universe Travel.

Using a complex form of in-line series-connected copper and silver caged crystals in my Morgan Antenna Tuner and Bucher QRM Local Waveform Suppressor before the signal gets to my Dunwoody or Litz Crystal Detector circuit, I can easily induce the Theta Waveform in the human biological brain. Theta Brainwaves are ideal for "Waking Dreaming" or "Astral Travel" provided you can induce this at the very outset of deep sleep, to minimize the Blackout Effect of the low Delta Brainwave State.

In my early experiments with brainwave research along with Dr. Hershel Toomim and Dr. John C. Lilly, we determined that the Theta State could be induced with some difficulty using biofeedback devices. My Parallel World Induction Device eliminates the necessity for a biofeedback machine.

The net effect is that my Parallel Universe Induction Device acts on anyone within the operating parameter of the device and they are within seconds swept rapidly and consciously from one universe to another without breakup of Stream-of-Consciousness.

What this means to you is that you can turn on the device and within a matter of moments or minutes you will be able to see, hear and feel the "Between-Lives" Interdimensional or Astral, Non-Organic Body, without a single, even momentary, break in consciousness caused by Delta Stage Blackout.

In short, your awareness or consciousness will be able to easily and without fear be able to travel between dimensions, remembering all the while that you are Astral Traveling Out of Body between a variety of "servers" or "dimensions".

The system operates independently of radio signals coming from local AM stations and, in fact, you will have to learn how to tune them out using what is called "Nulling" on the Bucher, Morgan and Dunwoody devices; it is a process which must be thoroughly mastered in order for the Induction Device to perform.

In my Parallel World Training Center will be several such PW Induction Devices. You sit straight up in a comfortable chair, eyes open or closed, and get sounds, images, feelings, sensations and memories from other dimensions in which you have traveled or are currently existing; a sort of "Past Lives Induction" that gives you free and easy access to memories and impressions of your many other lives in Parallel Universes and Time Frames as they happened, will happen or are happening right now.

In the PW Induction Training, you actually create a Log Book of Past and Parallel Lifetimes and events. Your Bardo Guide will turn on the Induction Device and, sitting within its operating field, you will be easily able to see, hear, feel and describe events which you have experienced over your Essential Self timeline which is almost infinite.

Your Parallel World Log Book is an incredible tool for many things; revived long-lost skills, improved memory and attention, expansion of consciousness and awareness and a sense of existence far above and beyond space, time and organic life.

There's more, much more. I'll be demonstrating my new Parallel World Induction Device at the Easter Workshop, and I'm sure you will be as excited as I am about this incredible new Work Tool!!!

Beta-Blocker Amulet is also attractive jewelry!

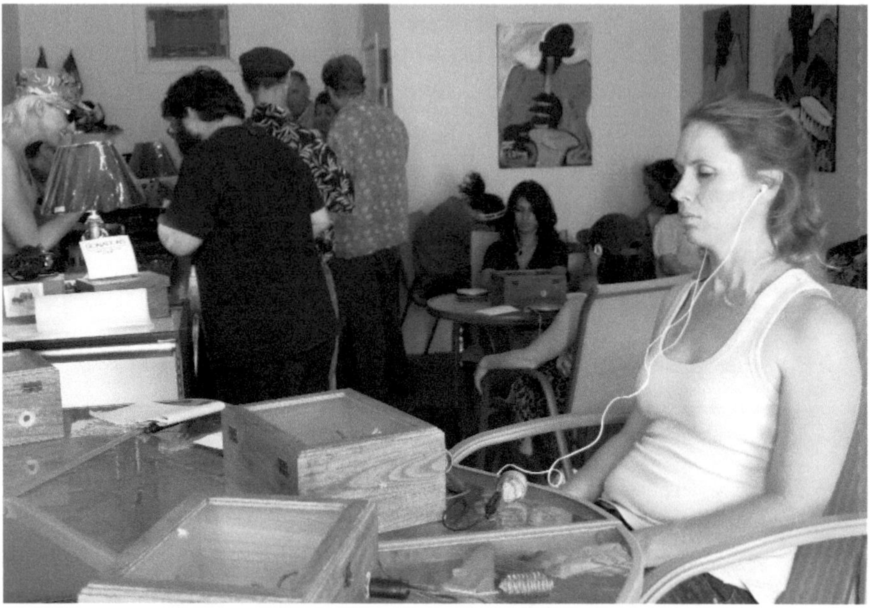

Slipstream Theory in practice, Neal St. Center.

Practical application of Slipstream Theory results in Parallel World Soul-Retrieval Workshop at the Neal St. Center.

Chapter Seven

SLIPSTREAM THEORY

A slipstream in time/space discontinuums operates in much the same way as a car following an 18 wheeler truck in a heavy rainstorm. The car travels forward at an increased rate with less energy required for the motion, riding on the break effect of the vehicle ahead.

In Interdimensional Slipstreaming the effect is produced by a subtle field of oscillation/resonance utilizing the regenerative heterodyne principle found only in a certain type of vintage crystal radio, discovered about 1908-1912 by several early communications pioneers such as Dunwoody and Litz.

The radio signals produced do nothing in and of themselves. Their frequencies and wavelengths are all wrong, but . . .

When tuning through the dial of the AM band — Amplitude Modulation — one discovers that a broadcast transmission from a single station is very likely to be found in several places along the broadcast band, within the range of the broadcast spectrum.

These are quite simply reflections of the transmission, occurring in multiples in the same way that a double-rainbow occurs, which is to say, in very predictable patterns of resonance, meaning that waves resonate and make sub-waves which are larger and smaller and which occur in definite intervals along the electromagnetic spectrum.

We can easily determine the validity of this in sound waves; musicians are very aware of sub-tonics, harmonics and other phenomena, particularly in the case of stringed instruments. Guitarists will often tune by harmonics on the high end of the neck near the body of the guitar.

Harmonics were first documented by the ancient Hellenic Greek philosopher/scientist Pythagoras in his study of the effects of bridged string instruments, leading to his theories of mathematics.

One can see these same effects in the visible light spectrum using special forms of prismatic lenses.

Harmonics occur when a wave has been propagated. Self-propagating waves cause sub-waves and sometimes superwaves, resulting in such oceanic phenomena as tsunami.

Waves tend to crash against one another, subtracting parts of their wave functions mutually, creating sub-waves and sub-tonic waves. These resultant "diminished" waveforms have "jiggle" and "bounce" among other sub-tonic wave phenomena.

Along with this we must consider beat-frequency oscillation and nulling effects on a waveform. Beat frequency oscillation reduces the effect of a wave, but not the wave itself, and results in a nulling effect in which the wave is no longer detectable in the output stage of reception and/or reproduction, as in the case of audio waves coming through headphones or a speaker-reproducer.

The set-up of waves can be modified by oscillation, resonance and nulling, producing sub-sets of waves both larger and smaller in predictable patterns of recurrence along the EMS (electromagnetic spectrum).

Along with this resultant chordal effect of combinant and interrupted waveforms, we can introduce the resonance effect, which can be demonstrated by the use of tuning forks; the commonest experiment in this area is the striking of one tuning fork which is then placed near, but not in direct contact with, another tuning fork of similar frequency.

The same effect can be obtained, that of vibrating the second tuning fork with the nearby vibrations of the struck tuning fork, when the second tuning fork has a frequency rating which is harmonic with the first, which may occur in a variety of fractionary equivalencies.

The transmission of soundwaves through water is profoundly affected by the liquid medium, as opposed to air medium. Space does not transmit soundwaves — although radio, x-ray, cosmic ray and lightwaves can travel through a vacuum — as well as many seen and unseen, known and unknown to present-day science waves can do.

The human brain thinks, reasons, rationalizes and compares in the associative range, between 14-30 CPS, or Cycles Per Second.

During sleep, the brain reduces its frequency down to about 4 CPS in the low Delta Region, the Realm of the Undifferentiated Universal Consciousness, within which the Ego is totally temporarily annihilated, then in sleep, it typically scales back up into the Alpha Range at about 8-13 CPS, the Dreaming Realm.

At the optimum operating range of 4-7 CPS it enters into the Intuitive or Telepathic-Psychic Range, within which mediums, psychics, remote viewers and telepaths typically can function with a low sense of Ego and a high sense of Cosmic Awareness.

During the 1960s I operated as a Remote Viewer, both as a civilian and in the military. From 1972 until 1977 I assisted my friends, doctors John C. Lilly and Hershel Toomim, at their request, in demonstrating the generation of the Theta State at volitionary will in lab tests during the development of early Biofeedback devices.

In the development of my Spirit XTAL Radio, I happened to stumble upon a phenomenon; I noticed that, when the crystal radio was tuned in a specific pattern of nulling using a combination of special quartz and black tourmaline array antenna loads, Bucher QRM rejector, Morgan Coil Antenna Tuner and Dunwoody Detector along with a unique high-impedance audio reproducer, the Theta Effect was generated without effort on my part to suppress Beta and Alpha; the field-effect of the system did all the work, obviating the necessity for a self-induced Theta State.

My assumption at this time is that somehow the field effect of the crystal radio array along with the crystal antenna load must have set up a harmonic field that caused the water in the braincases within the field to resonate at about 7 CPS, the high Theta Range, thus reproducing the Theta Psychic Effect and breaking down to some degree the "Branes" between Parallel Worlds.

The net effect was an instant break in reality-blockages, making several realities/Parallel Worlds available at the same time with no leakage or break in the thread of consciousness, just as one would experience in a self-induced or other-induced hypnagogic state. In short, the walls between the worlds suddenly diminished slightly, enough to "see" the other worlds quite easily and to report the images, sounds, feelings, sensations and knowingnesses of those alternate universe realities in much the same way that a psychic medium would do, without the necessity of mastering the psychic's "gift" powers.

I subsequently hooked a number of experimental volunteers to the unit and found that they were also able to do the same, without any prior experience or skill level at psychic remote reading or spirit mediumship.

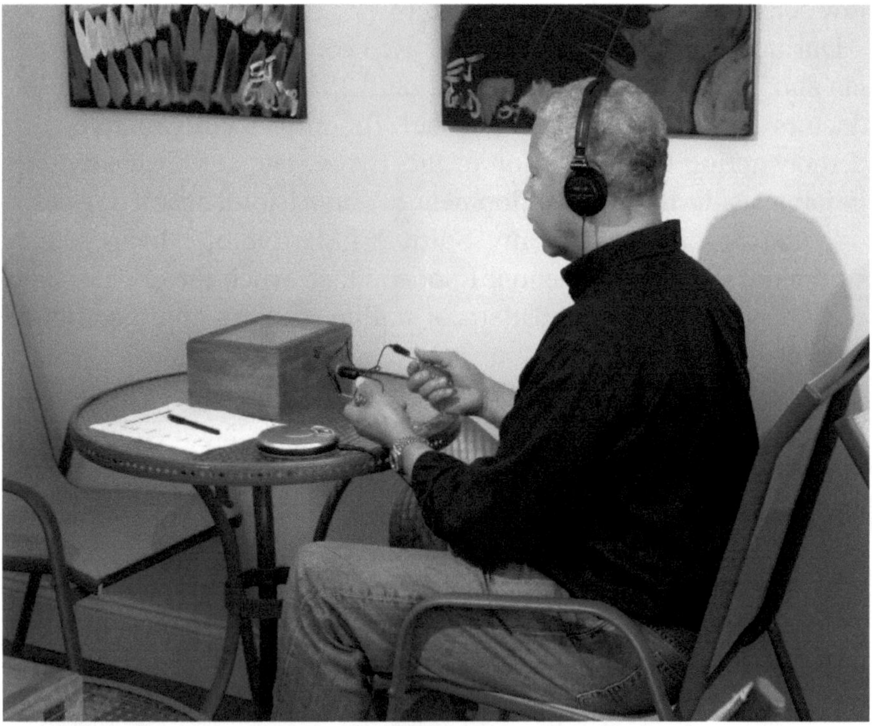

Experimental volunteers tested the unit.

Chapter Eight

WHAT IS JAUNTING?

The Stars My Destination by Alfred Bester

Story review and synopsis co-authored
by Claude Needham and E.J. Gold, editors of Galaxyezine.

Set in the future. Jaunting has become commonplace — the ability to teleport from one known position to another memorized position. This has dramatically changed the dwelling patterns, security patterns, and class structure of humanity on planet Earth. As the story opens, the absolute upper limit for a jaunt is 1000 miles and jaunting through space is not possible, but rocket travel throughout the solar system is commonplace, financed through ruthless economic exploitation.

At the moment the Outer and Inner planets and moons are at war, and the Outer Worlds seem to have the upper hand.

Gully Foyle is our main protagonist. There are several parallel plots with strong characters, and four female characters play key roles in the development of the Gully Foyle storyline.

The book opens with Gully Foyle surviving by wits, stamina and sheer stubborn will-to-live. We find him at the outset somehow aboard the crippled spaceship "Nomad." He has spent one hundred and seventy days living in a coffin-sized tool room, the only airtight space left on the damaged ship; his days are punctuated only by quick forays into the main part of the wrecked freighter spacecraft to collect a bit of food and air.

Gulliver Foyle, Mechanic's Mate 3rd Class, thirty years old, big boned, and tough — "too easy for trouble, too slow for fun, too empty for friendship, too lazy for love. A lethargic, dumb untrained ox of a man — sluggish and indifferent." A man of physical strength and intellectual potential stunted by lack of ambition. Energies at minimum. The stereotype Common Man. Some unexpected shock might possibly awaken him, but Psych Tech Division cannot find the key. Not recommended for promotion. Has reached a dead end. Gully Foyle spends his days waiting and hoping for rescue.

Then a ship comes, the Vorga. At first he is convinced it is only hallucination. He fires the rescue flares. The approaching ship slows in response to his flare, then speeds along its course sunward — leaving him to die. This is the single galvanizing event of his life. Now he has a reason to live, a reason to learn, a reason to not be lazy, a reason to do and be anything — he is going to exact his revenge on the captain and crew of the Vorga.

He fixes the ship in rather ingenious ways, and somehow limps sunward toward the inner planets which include Earth. He is now headed toward a rendezvous with a destiny of revenge that is now etched into the fabric of space and time, and he will not be denied his revenge. All the potential intelligence that has lain dormant under a cap of brute stupidity is now moving in an inevitable course of vengeance.

The wrecked ship ever so slowly takes Gully Foyle into the asteroid belt, where he is picked up by the Scientific Race — the descendants of a research team that had been lost and abandoned in the shattered sub-planetary masses between Mars and Jupiter. They are the only remaining humanoid savages of the 24th century.

Gully is given a wife, Moira, the first of many to come, as genetic distribution is vital in this isolated colony of savages. More importantly for the flow of the novel, he is given a full quite horrible facial tattoo. We never see the tattoo in full description. It is compared to a Maori facial design. Most of what we know of the tattoo comes from the reaction of others. With only one exception (much later in the book) everyone reacts with complete and utter disgust, horror, revulsion and/or terror at his facial mask. Basically the facial mask is an exact representation of the brutish feral core of his being bent on revenge of the coarsest kind. His bestial drive for revenge is comparable to the raw energy of a supernova turned-in upon itself.

Gully Foyle finds a serviceable weekender space yacht that has been patched into the woven steel fabric of the Scientific Race's floating tent city — made from ships and mining junk collected over the decades from the asteroid belt. With zero regard for the lives and safety of the savages who rescued him, Foyle blasts off, ripping the little weekender cruise space vessel from the side of the station, leaving a wrecked encampment behind him, after telling Moira to warn everyone to run into safety, just after he slugs her and throws her from the ship and blasts off, leaving a giant hole in the side of the space colony.

Moira is the first and last woman who will play an important part in Gully Foyle's adventure. She is intelligent, quick and compliant. She is played like the Tahitian native women in Mutiny on the Bounty. In fact, put that whole Tahitian village into space, give them ritualized slang and behavior powerfully degenerated from scientific heritage, and you have the "Scientific Race" in a nutshell.

Next we find Gully Foyle living in a hospital undergoing jaunt rehabilitation under the tutelage of Robin Wednesbury. She is a good teacher and good-hearted tall, lovely woman of African descent, brilliant and cultivated, but handicapped by the fact that she is a "telesend," a one-way telepath who can send but not receive; on a good day, with extreme concentration, she is sometimes able to selectively send to an individual; on bad days with low concentration or under stress she might send to everyone within telesend distance. In a world in which every telepath was documented and scheduled for employment at least 10 years into the future, she is totally ignored and not even registered. There is in Bester's World zero value for a telesend but there is very high regard for extraction of secrets and intent from a social, political or business target. However useless to the exploitation of business and political secrets her talent may have been, it was ideal for teaching, and teach she did; she directed classes in "Remedial Jaunting" for medically-challenged patients.

Jaunting can be interrupted by a specific form of concussion, drugs that disrupt concentration and confusion, in the same way that a patient might need physical rehabilitation to walk after surgery or an accident, but in the case of jaunting it made the difference between being a real person and being a sub-human to be exploited in the grossest fashion. Non-jaunters had little life in this new culture, in which it was considered the height of cool for an aristocrat to walk from a room rather than jaunt; a non-jaunter was considered unemployable except for the bacteria mines.

Gully Foyle does not actually need jaunt rehabilitation. He is malingering as a place to be while he investigates where the ship Vorga and her crew may be. When Robin discovers his malingering she threatens to expose him. He in turn threatens to have her turned over to the military police as an alien-belligerent because she was caught planetside with her sister and mother left on Callisto when the war started. He uses threats of death at the hands of the military interrogators to keep her quiet, and rapes her in response to her sudden revulsion at his face and soul.

We now transition to the next scene; the Presteign household, a massive economic power run by the family potentate Presteign of Presteign. They are looking for the wrecked space ship Nomad because it was carrying a cargo of 20 million credits of platinum and 20 pounds of PyrE. The PyrE is a psychokinetically induced thermonuclear explosive of the highest order. Twenty pounds might turn the earth into a droplet of molten slag and make the sun go nova. They want it back so they can win the war with the Outer Planets, and Presteign knows he could hold it for ransom from the military and make an even larger fortune and have more power for himself.

A new ship is being christened in the Presteign commercial fleet. Coincidentally, Foyle is there, attempting to blow up the Vorga which is also docked in the same space field. He does not succeed and is captured; he is questioned by the best interrogation money can buy, but he does not crack. The hatred and burning desire for revenge is so deep that under drugs and torture he reveals nothing but his desire to find Vorga. Since telepaths are booked solid and they don't want the telepaths to get wind of the PyrE's existence, they decide to put Gully into Gouffre Martel — the deepest darkest hole of a prison on the planet. They expect this to break him.

Gouffre Martel is dark and twisted to prevent jaunting. One must know the exact location to which one is to jaunt. In the depths of the dark maze which is Gouffre Martel, one can't calculate destination points, hence one can't jaunt. Those that get desperate and try to jaunt without knowing where they are, do what is called a "blue jaunt." The result of a blue jaunt is an explosion as the jaunter materializes in the solid stone into which the caverns of Gouffre Martel are carved. No one has ever escaped from Gouffre Martel.

It is here in the depths of Gouffre Martel that Foyle encounters the third strong female protagonist in the story, Jizbella McQueen. She is hot-tempered, independent, intelligent, and she is serving five years of "cure" in Gouffre Martel for larceny. She is a strong independent woman in a culture where women are relegated to the worst forms of paternal, fraternal, and husband protection. Women are kept as social sexual and service creatures, like pets, but not Jiz. She is strong and she is in total revolt against 24th century society.

Time served in Gouffre Martel is in total isolation, a completely enforced solitary confinement, but they found what is called a "whisper lane." In the caverns the accidents of rock can make passages where a whisper can be carried for thousands of yards. They find such a lane and proceeded to become friends and plot an escape. Jiz began the process of educating Gully. Up until this point in the book his language was coarse gutter speak. Now he learns a bit of other language so that on the Outside, Gully can have a greater chance for his revenge. But this momentary human relationship is like a friendship between a fox and hen. It will last only until the fox is hungry.

Foyle's plans for eventual escape are destroyed when he is interviewed by Dagenham, an agent for Presteign. The military, the corporations, the war alliances all of these agencies are so caught up in their own intrigues that they can't recognize the simplicity of Foyle's revenge motivation. They each see him as a spy, working for the other side. They are Machiavellian to the max, and he is equally vengeful and obsessive. During the interview, more information is divulged to Foyle then they intend; they let slip the fact that aboard ship were 20 million credits and something even more important, so important and so secret that they can't hire a telepath to extract it from him; he fights his way out, locates Jiz, and they both escape into the night; they make love, and become partners.

Next we see Jiz getting Gully a face job. Tattoos are non-existent in the 24th century, a remnant of savagery and stone age civilization. Jiz doesn't want his face surgically replaced entirely, so they opt for tattoo removal using oxidizing acids. The operation is extremely painful. At first Jiz withholds the extra 1000 credits necessary for the anaesthetic. Here we get a hint of the toll that Gully's brutish face and nature are having on Jiz. She loves him, and yet she wants to punish him for his brutality. We are given a description of the interesting operating-theatre. It seems the doctor is a collector of natural medical freaks; your imagination can run wild as to the variety of human-based specima he houses in his "zoo."

Partway through the operation, Jiz gives in, and pays the extra thousand credits for the anaesthetic. Just as the operation is completed, the corporate militia attack, killing a compatriot of Jiz, and forcing the two to flee. Gully steals the keys to a Saturn Weekender from the dead body of Jiz's old friend and they proceed to collect the 20 million credits in platinum from the wrecked Nomad.

At this point the Love/Hate nature of Jiz's feelings for Gully comes to a head. She continues to work with him "for the duration" of the collection of the platinum from the wrecked Nomad; 20 million credits is in the 24th century equivalent to 20 billion dollars in today's economy.

During the trip from Earth to the wreck of the Nomad, Jiz makes Gully mad, thrusts a mirror in his face, showing him the tattoo marks emblazoned in red across it, and informs Gully of a characteristic that will underscore his actions throughout the rest of the story: "You said you wished you could carry me in your pocket to stick pins in you when you lose control. You've got your face . . . you can't ever lose control, Gully. You'll never be able to drink too much, eat too much, love too much, hate too much . . . you'll have to hold yourself in an iron grip." Gully had a brilliant reddish "stigmata" burned so deep into his facial muscle that, without removing his face down to the bone and replacing it with another, he was stuck with his "little reminding factor."

While Gully and Jiz are going about the business of grabbing the safe and trying to open it, the corporate ships from Presteign arrive. Under the threat of attack they rip the safe from the wall of the Nomad and blast it toward the cargo bay of the Weekender like "catching a baseball in your cap." The safe jams in the hatchway making it impossible for Jiz to board the ship. Not willing to give up his chance at revenge for anything, Gully blasts off leaving Jiz to face the forces from Presteign. End of this segment of the serialized Galaxy SF novel.

A year has passed. The war has gained momentum and has grown rapidly from a distant romantic affair somewhere very far away to an ever-present poisoning effect on every aspect of life on Earth. Spy scares, draft-jaunting and labor jaunting became critical problems. The hysterical have become informers and lynch mobs. An ominous foreboding paralyzes every home, in which people are "turtled". All of this is enlivened only by the advent of "The Four Mile Circus".

"The Four Mile Circus" is the nickname for the grotesque entourage of Geoffrey Fourmyle of Ceres. Geoffrey is the classic "nouveau riche" of all time — brash, enormously wealthy, with a talent for squandering outlandish sums of money on everything possibly amusing to break his utter boredom. He maintains an entourage of circus-like proportions — dancers, troubadours, cooks, maintenance teams, engineers, actors, musicians, artists, flakes, bakers and fakirs. In addition to all of the usual frivolity, Geoffrey maintains a traveling library of the first order, a college, an on-staff team of inventors and a whole research and development team, all for the single-minded purpose of carrying out his revenge, for "Geoffrey Fourmyle" is, in fact, Gully Foyle.

In order to break into society, Gully rescues, kidnaps and convinces Robin Wednesbury that it will be to their mutual benefit for her to use her telesend capabilities to stuff into his brain the correct polite conversation for moving within society, for Gully has determined that to uncover the next lead in his quest for revenge, he must move in High Society, and he knows that he cannot learn quickly enough to fake his way through it, so he plans to be the ventriloquist dummy to Robin.

Gully now has an inexhaustible supply of money. He is also holding the secret to the location of the solar system's most formidable explosive substance. He has been transformed by surgeons of the Mars Commando Brigade into the solar system's most formidable fighting machine, an operation normally reserved for the Best of the Best military special services; they have rewired every nerve plexus. Microscopic transistors and transformers are buried in muscle and bone. A miniature platinum outlet connects his web of fighting equipment to a powerpack. He has control switches embedded in his teeth to help control his emotions, to control his time sense and his body's response-time and strength. He is able to speed up time to the degree that he can waltz around foes as if they are statues. In addition, Gully has one huge Achilles heel; if he becomes excited in any way — anger, sexual tension, fear . . . anything — his blood will cause the scars of his removed facial tattoo to become visible, thus giving away who he is, and destroying any hope for revenge against the Vorga. His scientists have calculated statistically that he has just three months before his charade is discovered and he is found out, so he must work quickly within this time-limit which, by the way, was one suggestion from Horace Gold, the editor of Galaxy SF magazine, within which this serialized novel first appeared.

Robin works with Gully partially as prisoner, partially as co-conspirator — she is trying to find her sister and mother — partially as a social tutor, and unbeknownst to Gully, as an unwitting and unwilling admirer. There is something within Gully, not charm, not grace, not respect, not anything that we can put our finger on, that wins women over to him despite themselves and his basic raw rage. These are not women looking for rough-trade, or a walk on the wildside. It is not their will to self-destruction. There is something else that brings them to love him. Unfortunately he is such an absolute brute in his unwavering quest for vengeance that he throws them all aside without ever noticing that they cared about him. Everything that Gully does is calculated toward a single purpose and, as far as we can tell, he will walk over and destroy anyone in order to get his revenge. It is only his reflection in the women around him that give any hint that he might have a redeeming quality and each of them despise him in addition to anything softer they may feel. Ultimately they will all betray him and contribute to his resurrection . . .

It is in the next section of the serialized novel that Gully uses his wits, money and special body augmentation to travel from one adventure to the next, from one planet to the next, following his leads like a dime-novel detective does; by grit and heavy-handed savagery.

Because of a profound post-hypnotic implant, his leads give up secrets and then die by heart-stoppage. Gully performs open-heart surgery on one crew member of the Vorga, who looks down at his chest gaping open, his heart pumping on top of the side table. Gully questions him about who was responsible for ordering his being left to rot in space; he is told that the only one who might know is the captain of the Vorga, who is now a member of the Skoptsy Colony. The Skoptsy have had all their sensory input cut away; they live totally within their own heads with no input from the outside world, so Gully steals a telepath to get at this last clue.

Meanwhile Gully has fallen in love with Olivia Presteign, the other-sighted daughter of his arch-nemesis, Presteign of Presteign. Olivia can see in the infrared and microwave, but not the visible spectrum, which means she is not able to jaunt. Being jaunt-challenged in this world of the future is the absolute worst disability. It is when Gully asks Robin to help him win Olivia's heart that we see through the deep wound this causes that she — who has been raped and mistreated most by Gully in his revenge quest — has come to love him also. Meanwhile, Gully is visited at key pivotal plot-points by a mysterious Burning Man, who flickers in and out of existence like a candle flame, blazing upon each moment of significant history for Gully.

Using the telepath, a pathetic little 70 year old baby, to get at the Skoptsy captain of the Vorga, Gully finds out that the villain who decided to leave him to rot in space is the current love of his life,

Olivia Presteign. The Presteign family are evidently all cursed by a common thirst for blood and pirating. Olivia in a secret venture to quench this thirst on the vile side of her nature, has been running refugees from the outer planets to Earth, except that she takes their money, then dumps them out of the ship naked into space to blow up from their own internal pressure, which is precisely what she was busy doing with Robin's sister and mother when they came across the Vorga.

Here Gully is hit by two strong shocks. The charming and beautiful Olivia Presteign turns out to be the monster upon whom he has been seeking revenge, and he now understands that even had he been picked up, he would have been scuttled into space to explode, blue and naked. He was simply in the wrong place at the wrong time. In one blinding moment of revelation, the Light at the End of his Tunnel has been ripped away. Gully Foyle is now a character ready for destruction in a pool of self-pity, or someone who is finally ready for a major transformation as risen dough is ready to be baked into a loaf of bread.

It is in this state that Gully goes to an attorney to negotiate the terms of his surrender to the authorities. He wants to give himself up. He wants to be purged. He wants to be punished for what he has done and made clean again. But . . . his attorney has been transformed into a fighting machine to kidnap Gully for the Outer Planets. They have been tracking him for years, waiting for a chance to grab him for two reasons; they want the PyrE explosive, and they want to cut the secret of space-jaunting from him; so far, he is the only human who has achieved jaunting in space.

It turns out that the Nomad was wrecked by raiders from the Outer Planets. They grabbed Gully, the lone survivor, and put him adrift in space in his space suit with a transponder signaling for help, a space-suited lure for any would-be rescue ship, the idea being to jump and destroy any rescuing ship. Gully was poisoned bait who spelled doom to anyone who stopped to help; a live, wriggling worm on a deadly hook.

In order to make their little trap work, the bad guys had to take Gully 600,000 miles closer to the sun, into the standard shipping lanes, otherwise no one would have seen him and attempted a rescue. Well, Gully surprised his captors by space-jaunting back to the Nomad — the first human to jaunt through the voidness of space. This the Outer Planets wanted to know how . . . with this secret, they could win the war and totally control the Solar System.

So not only was Gully's rescue ill-fated by luck; had he been rescued, it would have been the ruin of anyone making the attempt. At this point, all his possible anger at others for not stopping to help him suddenly drains away. He is knocked out, drugged and dragged to St. Peter's Old Cathedral — the most recent home of the Four Mile Circus, where the attorney/secret agent expects to find the PyrE before leaving with his captured prize to the Outer Planets.

Presteign has at this point decided to explode the bits and pieces of PyrE left around by Gully during his research team's efforts to uncover its secrets in an attempt to locate the other 18 pounds of PyrE that remain hidden. This explosion of dust particles is sufficient to devastate St. Peters and — you guessed it — to ignite Gully like a candle flame.

It has been Gully Foyle jaunting throughout time that caused the Burning Man sightings throughout the story; in the explosion, Gully's senses are all mixed up, but he receives telepathic instructions from Robin during a future appearance of the burning man that tells him to climb from the wreckage of St. Peters to safety, which he does, just in time.

When his senses uncross in the Star Chamber of Castle Presteign, Gully is left with the task of sorting out what he will do with the PyrE, the location of which is known only by himself. The war effort wants it. Presteign wants it — maybe for Earth, maybe for the Outer Planets, the highest bidder. It is a malfunctioning robot that helps Gully sort out his intentions. He decides to give the secret of PyrE to the entire world. He distributes enough bars of the substance around the planet on a jaunting run, so that either the military is forced to divulge to the public what they have, so they will turn it in or risk accidental explosion of the whole planet. This forces the arrogant aristocracy to bring the people in on the secret that threatens their very existence. A rather democratic move reminiscent of the optimistic egalitarian 1950s, during which this serialized novel, later published in book form, was written by an equally optimistic and egalitarian Alfred Bester.

Gully does his best to break himself from his past and then jaunts to the Scientific Colony to Moira and the Scientific Peoples. Here he enters into a cocoon-like womb of delirium — a vision quest from which he may exit a transformed man.

Alfie Bester and E.J. Gold discuss writing and word processors, 1983.

Chapter Nine

BETA-BLOCKER TECHNICALS

Theta isn't "not there" and then it "is." Theta is always active, as is Alpha, but they are dominated by an overwhelming force, which we call Beta. Beta Brainwaves exist at the 14-28 Cycles Per Second (CPS) range within the electromagnetic spectrum (EMS).

Our target Operant Brainwave Range is between 8-14 CPS, ideally at around 8 CPS.

Combinant waveforms can involve Delta Brainwaves operating at around 4-6 CPS. These resultant secondary waveforms cause ripple and jiggle Moire patterns which produce a visionary state above the causative state, in turn producing cognitions in the revelatory sense.

Cosmic Cognitions in the psychic-intuitive state are understood by one in the Alpha-Theta State, but are often not interpretable in the Beta State, which is the state of the comparative/associative mind.

Thinking by comparison and association in the Beta-Brain is not the ideal condition for understanding concepts understood and appreciated in the Theta Mind.

One may not grasp the fundamentals of Theta-Mind concepts when in the grip of Beta, but in the psychic-intuitive functioning of Theta-Mind, these concepts become perfectly clear.

When in the presence of a Beta-Blocker, the experiencer must be open to the influence of the Beta-Blocker and receptive to the Theta-Mind State. If one is heavily immersed in the Organic Body, organic habits may dominate the Theta experience for a while, but eventually this will break down under continued exposure to the Beta-Blocker.

There are some conditions of heavy Beta Domination and Beta-Dependency that make it difficult for some individuals to allow the Beta-Block effect for some time. A few very Beta-Dependent individuals will actively and frantically fight the Beta-Block in order to remain in the Beta-Brain.

For deeply dedicated individuals at this level of deep Beta-Dependency, a stronger Beta-Block is indicated, such as the BB-102 and the BB-110 which are hyper-heterodyne active transmitters with powerful signal-to-noise ratios in the 1440 KiloHertz range.

Additional breakdown of Beta can be achieved with the addition of audio frequency waveforms through white-noise emitters and other thought-processing blockage devices acting in concert with the radio-frequency fluid resonance brainwave-breaking patterns of the Beta-Blocker.

Low-level brainwaves in the Alpha-Theta-Delta range can be influenced by very low level transmissions in the milli-microvolt range, and synaptic patterns of comparative thought can be influenced well within the picofarad capacitance potential range of arc, therefore a weak radio signal can easily influence the salt-water computer without penetration of the braincase merely by induction.

Induction, resonance and oscillation are the three basic phase factors which produce the Beta-Block Wave-Generation Field-Effect. The Beta-Blocker is a Field-Effect Generator (FEG) of the lowest order but its Resonance-Effect is not boundaried by distance. The FEG operates on the Einsteinian

Quantum Effect, "Spooky Action At A Distance" covered in his Special Field Theory within the spectrum of the Unified Field Theory.

Using String Theory Concept, The String is a single strand of multiple interconnectedness at definite and predictable nodes, creating a single field of multiple dimensions. Those multiple dimensions are the causative factor which produces "Phasing", a result of "jiggle", which in turn is produced by imperfect phasing of waveforms.

Phasing of Membranes, the boundaries of reality-forms, is the causative factor of "Sparking," the Primum Mobile of an infinite series of Big Bangs, producing in turn a large nearly infinite number of resultant explosions of space, time, energy and matter which are commonly called "universes."

Each universe is cross-connected with all other universes by a commonality of Phase Factors and Resonant Tones within the limits of the EMS, but the physics of each universe may be very different when taken in relation to one another.

This relativity is the basis of universal "cross-talk" and "chatter" which results in a "ghosting" effect of the Universal Mind, creating what appears to be a multiplicity of beings and entities, all stemming from a single source.

Cross-talk and Chatter can be interpreted as communication between Beings.

Voyaging between Alternate Dimensions, or Parallel Worlds, is passive and requires no action of the individual; recognition of Voyaging is dependent upon Waking State levels of Attention in the Psychic-Intuitive State. The optimum level of existence of a Being is to be aware of one's continual shifting from dimension to dimension at intervals of several seconds to several hours.

This Phase-Shifting Effect which causes the Eternal Voyager to constantly and unendingly penetrate the Membranes (Branes) of an almost-infinite number of Bubble Domains can be easily seen when in the Theta State.

This awareness can help the Being assimilate and transcend all organic experiences and phenomena, thus freeing the Being from attachment to the organic while remaining in a state of high cosmic attention and indifference to outcome or temporary condition.

What that means is a way "Off the Wheel" while remaining within the boundaries of the Wheel, to be "in the world yet not of it," a desirable state for any Being, allowing one to operate within the sphere of world phenomena without being dependent upon phenomena.

This process can be used to achieve the Walking Waking State, which is often referred to as "dying before you die".

In another universe, JFK is still alive and well . . .

In a Parallel World, nuclear war happened decades ago.

Chapter Ten

STRING THEORY

String Theory is a single pixel creating a woven trigonometrical pattern in a single instant. The intersections are "nodes" or "locations." It intersects itself at apparently random but actually predictable intersections, nodes, or locations. In Theta, this vision of the universe is quite accessible.

The composite whole could be called a multi-dimensional weaving, creating co-existent levels separated by voidness. Norton Street contains all levels and could be called a megalocosmos. Cosmo Street exists as similar levels in an almost infinite array within each object, or model, in Norton Street. The nature of objects in Norton Street is such that a shelf unit full of books is in fact a single polygonic model containing a nearly infinite array of Cosmo Street orbs.

Each Cosmo Street contains a nearly infinite array of Bardotowns.

Each Bardotown contains a definite finite number of Nimbus or Rebirth Stations.

Each Rebirth Station is strapped to an infinite number of Home Worlds.

Each Home World is strapped to a nearly infinite number of inhabitants or Avatars.

Each Avatar is connected simultaneously with all other Avatars, thus forming a Ground Plane in relation to Cosmo Street as the Antenna Plane.

The differential between dark and light energy creates the apparency of space, time and matter.

Time exists only in relation to curvature of space, creating magnetism, gravity — which in turn create the effects of magnetism, electricity and the electromagnetic spectrum.

The laws of physics are very different at each of these levels. Avatars at the level of Cosmo Street and Norton Street are composed of a nearly infinite number of what appear to be stars with space or voidness as the containing medium. They are boundaried by self-luminous skins, each one of which forms a permanent and eternal Body of Habits.

There is a direct correlation between these giant Body-of-Habits Avatars and local Home World Avatars within the physical organic universe.

One tends to take rebirth in the Home World Avatar most similar to the Cosmo Avatar one returns to in the between-lives state.

Signed,
Archangel Zadkiel

Archangel Zadkiel is a Cosmo avatar. What we call angels and archangels are actually avatars in the Bardotown through Cosmo levels; and what we call god or gods would be avatar(s) on the Norton level, outside which is an inchoate creator whom we call the Absolute, who never takes form and who has no attributes whatever, not even voidness. This nonexistent entity is the causative factor for everything we know and can be aware of.

In this sense, the entire system of universes could be viewed as a giant, sizeless complex simplicity roughly equivalent to our entertainment center or Ipod.

Taken as a whole, this system of cosmoses, megacosmoses and megalocosmoses is often referred to by psychic intuitives experiencing the waking state as The Cosmic Joke. The laughter you hear at the highest level would be that of the Absolute, punctuated at various intervals by primal screams, coughs, grunts, snorts and snores.

The electric potential created by the differential between ground plane and antenna is the causative factor for the energy constant within all universes.

Time, space and matter are products of the interaction between the two types of energy: dark and light.

Jiggle, flutter and wow create variations in the boundaries between energy domains existing along the frequency spectrum from smallest to largest wavelengths.

Reality domains exist within the octavic boundaries within the EMS (electromagnetic spectrum) in a periodic table of repetition, translatable by phase factors of resonance and oscillation.

The intersecting point weaves itself into the cosmic fabric in an instant, but voyaging along the lines creates an apparency of time flow.

Intersections mean taking the right bus at the right time to the right destination in Parallel World L315a!

Radio waves contain transmissions from the Big Bang.

Working on a crystal radio receiver with 1902-1912 vintage components.

Chapter Eleven

RADIO WAVES

In the early 1940s, the Germans had sensitive radio direction-finders that could easily locate and pinpoint as artillery targets the local oscillators of superheterodyne portable receivers used by GIs in foxholes in the battlefields of "Fortress Europe", and so the Pentagon outlawed portable superheterodyne tube radios on the battlefront.

GIs would not and could not give up their news, entertainment, weather and of course the latest baseball, football and other sports scores and game highlights. They sat in the rain and snow in their foxholes fighting constant boredom.

The Nazi Propaganda Ministry operated powerful broadcast stations that targeted U.S. Forces with American music, such as the popular Glenn Miller, Artie Shaw, the Andrews Sisters and more and there was the German Sexpot, Axis Sally, who told them what was REALLY going on, what the Big Brass didn't want them to know!!! You could also listen to the equally powerful broadcasts of the Armed Forces Radio Service, which my Dad Horace once administrated as a Colonel in the South Pacific in 1944-45. In the Postwar Years, Armed Forces Radio aired X Minus One, my Dad's radio show featuring stories from the pages of Galaxy SF Magazine, which he founded and edited for its first decade.

Many GIs had been Ham Radio Operators and "Homebrew" radio builders back home, so they knew how to make non-heterodyne mostly undetectable radios out of materials other than crystals which could be used as radio wave detectors, such as Gillette Blue razor blades. They made headsets out of wire, screws and washers, cut mounting boards out of ammo and kitchen supply crates, forms were of course your ever-present standard toilet paper rolls, and pencils and safety pins could be had for a few cigarettes, the prize commodity on any battlefield. Plans for foxhole radios circulated as "scuttlebutt" and many soldiers made razor-blade radios even though they had no idea how and why they worked.

The fact that radio receivers also transmit radio waves is an important factor in the development of the Beta-Blocker Field-Effect Generator which suppresses significantly the Beta-Brain Functions, exposing the Alpha-Theta Psychic-Instinctive in the 8-14 Cycles Per Second range.

Theta enhancers are mis-named. Actually the Beta is there all the time; it never ceases to function until brain-death. Suppression of Beta brainwave activity reveals the Theta Psychic-Instinctive.

What I have been trying to develop is a small, portable unit that can act as a Beta-Suppressor. The small size does not limit its potential, because the 4-meter radius of Field-Effect is quite sufficient for the artist's or musician's studio, for the inventor's shed or for the quiet intimate space of a psychic medium's séance parlor or a tarot-reader's table.

It fits in quite nicely within the sphere of the Spirit Altar and is very useful in the invocation of non-embodied entities, and in Sacred Prayer, opening up the dimensions and softening the walls, the boundaries, between domains of Parallel Worlds.

This wall-softening effect is actually the penetration of the boundaried reality-domains by the Theta Psychic-Instinctive, and can greatly enhance conscious astral travel and Voyaging Dreams.

With the Beta-Blocker, the Voyager can easily see Beyond the Veil imposed by the Beta-Brain. Many psychic-instinctive uses are now available to the Interdimensional Voyager with the Beta-Blocker.

Chapter Twelve

USES FOR BETA-BLOCKERS

Folks have been calling and emailing and writing in asking how the Beta-Blocker can actually be used, by which I take them to mean, how can one employ it, deploy it and thoroughly exploit its benefits . . .

This is a good question, and one which I will answer at great length during Workshops — however, a short response is, I think, in order . . . .

I mentioned in several previous commentaries that I could see this being used in several locations:

1. Body Work studios, such as Shiatsu or Rolfing centers
2. Séance Parlors
3. Tarot Reading Kiosks or areas
4. ABD reading spaces
5. Artists' Studios
6. Recording Studios

(does not interfere with recording equipment)

7. Design Studios
8. Work Cubicles
9. High-Stress environments
10. Children's Rooms

Along with all those ideas came a phone call today from an art dealer. I suggested using the Beta-Blocker in the Closing Room (Viewing Room is what he called it, but we all know it's a closing room where you finalize the sale).

But speaking of a Viewing Room, how about placing one of these near the Recently Bereaved in a Mortuary?

Could be of benefit in a reading room or library or study hall or homeroom or where kids do their homework, to aid in concentration and focus.

How about in a classroom full of ADD kids? Used in a space with folks who have trouble focusing attention or suddenly ripping out into alternate realities, this might help make transitions and balance a bit easier.

Prisons? I doubt they'd even consider it, but what an idea — give the prisoners another reality to duck off into.

How about ashrams, meditation spaces, chapels and, using the high-powered model, even cathedrals?

I can see engineers, physicians, dentists, draftsmen, anyone with a serious need to concentrate and focus attention, using a Beta-Blocker in their workspace, and certainly anyone in a high-stress lifestyle would want one in every room in their house, and maybe even one in the car. I can make an automobile model if someone really wants it.

I can see miniaturizing this device and carrying it in the pocket or purse.

. . . And a Retail Space . . . ? Wow! You know the effect of Beta-Block if you've ever been pissed off at me or Lee Lozowick or Zalman Schachter-Shalomi or Claudio Naranjo or Tarthang Tulku Rinpoche or Swami Rudy . . . You march in there, all raging and ready with vituperous venom, and by the time you get within a few feet of any of us, you're blissed out and giggling with glee. That's the effect of being within effective range of a Living Beta-Block.

Think of the Beta-Block Device as a sort of miniature low-powered Portable Personal Guru for Home or Office, and you'll get it.

And what's more, for only $55, which covers the cost of making them; Claude and I do NOT charge for our labor, we do that as a service. We've made some videos on what it takes to make one of these little puppies, and I think you'll agree that our labor is intense and difficult and easily worth a hundred bucks, but we don't want your money. We just want you to use the Beta-Block and see how far out you can get with it.

Distant Galaxies? Piddle. Those are close neighbors compared to Parallel Universes. Get Out There and PLAY!!!

Godd Team makes alternate worlds every day.

E.J. Gold on gorebaggtv explains the Beta-Blocker, March, 2009.

First wave of Beta-Blockers ready for shipping, April, 2009.

Chapter Thirteen

GHOST-HUNTING WITH THE SPIRIT RADIO

Using the CHILL FACTOR, locate the exact HAUNTING POINT caused by the presence of the restless spirit. Most haunts are the result of the inability of the spirit to move on to the next plane of existence. This can be for several reasons; extreme attachment to the Earth-Plane, unfinished business, revenge on the living for some transgression, fear of moving on or letting go of Earth-Plane existence even as a disembodied spirit, or a quest for justice, wanting their fates to be known by the living before passing on, and in some cases, mere confusion and not knowing that they have passed from organic life into the spirit world.

In each of these cases, it's strictly up to the Ghost-Hunter to know what to do. The Spirit Radio Ghost Hunter does NOT do anything to stop hauntings, it simply indicates the area where the haunting is taking place, and then only to a SPIRIT SENSITIVE. The average person will not necessarily automatically benefit from its use, however, it is possible for someone untrained to be trained or to train themselves to become a spirit medium through the use of this or similar devices.

Long exposure to spirit hauntings and incarcerations will produce in the Spirit Medium a profound sensitivity to the CHILL-EFFECT, which is easily recognizable by any professional Spirit Medium or Psychic-Sensitive.

Merely walk the device around the haunted area until the exact spot is found, then begin your usual procedures to exorcise the spirit. In the case of poltergeist activity, there may be a HEAT indicator rather than the usual Cold-Effect. This is easily determined by experiment in several haunting areas and will quickly be recognized by the Spirit Medium after a few tries.

Many trapped spirits have no idea how to vibrate air to make sounds, but for some reason they are easily able to distort or alter in some way radio waves and other electronic waveforms.

Your Spirit Radio is actually both a receiver and a transmitter, although the transmission signal is very weak and will not carry far. During World War II, the Germans used this fact that the radio receiver is also a weak transmitter to pinpoint GI foxholes where the soldiers were using receivers in order to give locations to their artillery and air support.

Using the transmitter effect, merely talk to the spirit and, using the earpiece or the OPTIONAL add-on high-impedance vintage headset or high-impedance amplifier-speaker, listen for a reply. Your set is pre-tuned to a commonly used easily accessed RESONANT frequency that disembodied spirits seem able to tune into and distort into hearable whisper tones.

Communicating with the spirit, you can easily determine the cause of the haunting and help them decide to move on to the next plane of existence.

Spirit-Medium Effect:

Profoundly useful in hauntings, the Spirit Radio is even more effective in controlled professional LEGITIMATE Spirit Séances. The TRANSCEIVER EFFECT makes for a more relaxed and easy Séance without the need for trance-state transmission or channeling of the voice.

Where preferred, the Spirit Medium can channel the voice, using the radio signals as a prompt and in some cases as a slight amplification. Some spirits are simply not able to communicate well or clearly and this blur or slurring can be reduced by the use of your Spirit Radio. The amplification is an illusion; actually it is a slight clarification of the communication through the RESONANCE EFFECT where the higher vibration signal is duplicated on a lower level of the electromagnetic spectrum (EMS) and made somewhat more detectable and understandable to the professional medium.

Where a spirit is at a great distance, the DISTANCE EFFECT of being on a different Plane of Existence or Parallel World will be greatly reduced.

Flight 103 left quite a few disembodied spirits in a state of utter confusion.

Healer-Cleanser is used by healers and body-workers to discharge "gunk" from healing hands.

Healers from Tequiha Foundation learn how to use the cleanser.

Chapter Fourteen

HEALER CLEANSING

Cleansing-Charger:

To produce the Cleanser-Charger for Body Workers, I've wound the antenna and ground loops into circular form. When you pass the palms of your hands through them one at a time, you'll notice two definite effects: a detectable increase in heat around the palm and back of the hand, or a detectable icy-cold well-known by spirit mediums and ghost-hunters.

This device is probably useless to the lay person, but an experienced HEALER will quickly see what the device actually does to remove the gunk deposited from the Healing Passes of the hands. It's sort of a "wiping" effect on the hands when used as a cleanser to remove the vibrations resulting from hand-passes over the affected areas.

You'll feel what has been described by Professional Healers and Body Workers as a light scraping effect somewhat similar to a lightly scratched sandpaper over the back of the hand. This "WIPING EFFECT" has been noted by every pro who has tested the device, but there's more . . . .

Charging Effect:

The Charging Effect comes from a definite and profound feeling of warmth and a "pull" or what has been described as a magnetic or gravitic effect on the hand as it passes through the airspace between the antenna and ground loops.

Closing the eyes when making hand-passes through the loops the first few times will give the Healer a better understanding of what speed of hand-pass is best for them. Varying the speed of the hand-pass will vary the effect of the CHARGE-UP.

You will eventually find the exact hand-pass speed that works for you, and it may vary from client to client.

Body-Worker Effect:

Try a few hand-passes within the FIELD-EFFECT between the two loops before performing your Body Work on a client. Notice the effect. Repeat the hand-passes as necessary. Some Body Workers have reported a single pass works well, others have said that repeated hand-passes work well for them. Use your intuitive and instinctive centrums to determine for yourself which is best, and be prepared to see that change as you learn to use the device more effectively and efficiently.

Combination Effect:

Some Healers and Body Workers who have used this device say that they like to cleanse the hands periodically while working, and that what they do is to first get the CLEANSING EFFECT and then wait a few moments, then make their hand-passes through the loops to get the CHARGE-UP EFFECT.

Don't forget to do a final CLEANSING EFFECT after your Healing or Body Work!!!

Chapter Fifteen

BRAINWAVE FUNCTIONS & ENTRAINMENT

The scientific principle of entrainment utilizes resonance, which means that a higher frequency will have corresponding periodic "notes" all along the EMS. Our bodies are mostly water, which is a liquid that easily carries current and sustains waves. Our brains operate by resonance, beat-frequency and oscillation. Pulses of electrical potential ripple through the pool of the brain and nervous system in much the same way that currents circulate in a stream-fed and stream-drained active pond.

The brainwave range literally dominates and forces a precise state of mind and emotion. The four most common types of brainwave frequency ranges tend to be well-mapped, but keep in mind that the functions of brainwaves tend to overlap, with Beta dominant after it has been entrained by common education — reading, writing, arithmetic, history, language, verbal communication and more. The left hemisphere of the brain tends to be linear, intellectual, practical and semi-logical, with extreme emphasis on time and task orientation, while the right hemisphere tends to be non-linear, abstract, creative, apperceptive, abruptly instinctive and, from Mr. Spock's viewpoint, totally illogical.

The Corpus Callosum serves as a connective "wheatstone" bridge between both hemispheres. The Callosum can be strengthened and enlarged somewhat by certain types of mental exercise which were the main cause of the enlargement of Edgar Cayce's Callosum, making it far more capable of transmitting data between hemispheres, thus allowing them to work together to increase our internal median brain-modem which seems to operate at faster speeds simply by increasing its "bandwidth". So all these brainwave states have been examined and charted to find the respective mindstates that are experienced when they are present. We are learning more and more about how these states can be summoned and harnessed to create mindstates that are conducive to expanded states of consciousness. For example, the best moments of creativity, those Eureka! flashes, occur mostly when Theta waves are a majority. This is also when we are verging between waking and sleeping when we are in that Twilight Zone-like consciousness mindfully floating at the gateway of sleep.

Brainwave entrainment or "brainwave synchronization," causes brainwave frequency to correspond to a specific functional range. Entrainment is dependent on a natural phenomenon in which any organic salt-water computer, which we call "a brain" will automatically alter its dominant operating frequency range when guided by internal or external stimuli which can be auditory as in the case of binaural beats, visual, as with a dream-induction device such as the Russian SOMA, and of course direct or indirect electromagnetic radiation or induction.

In the standard form of binaural beat brainwave entrainment, two electromagnetic spectrum signals, which can be audio or radio range generate a beat frequency oscillation, something which will be familiar to all Ham Radio operators using CW (Continuous Wave) transmissions and Morse code keying.

The crash effect between the two waveforms produces a subtonic wave or harmonic frequency, which is the sum of the difference between the two primary or tonic frequencies and which can occur at several places along the EMS (electromagnetic spectrum).

A 1550 Hz tone against a 1557.8 Hz tone will produce a subsonic 7.8 Hz tone, one of the Primary Theta States, often called the "Schumann Resonance Effect", which can produce a directly transformative experience, with no audio signal or external clue whatever that something has taken place within the brain, other than the state induction.

Local field potentials (LFPs) using electrodes directly contacting the brain occur at different frequency ranges, in different brain areas, and some resonant oscillations have been related to particular behaviors.

A mind machine called a psychowalkman uses pulsing rhythmic sound and flashing light to alter the brainwave frequency. Some mind machines induce deep states of relaxation, concentration and in some cases altered states of consciousness which will be familiar states to the meditator and shaman. This mechanical method of achieving deep meditative or psychic states is called "Entrainment" by the mind-machine industry.

Entrainment Sessions typically aim at causing the brainwave frequency to adjust to Target Frequencies of Delta (1-3 Hertz), Theta (4-7 Hz), Alpha (8-12 Hz) by suppressing the normally dominant Beta (13-40 Hz).

Some respectable clinical research has been done on the use of auditory and visual stimulation to improve cognitive abilities in the learning-disabled.

.5 — Epsilon Range — high ecstatic states of consciousness, spiritual revelations, out-of-body experiences, Yogic "suspended animation", universal mind.

1-3 — Delta range — deep non-REM (Rapid Eye Movement) dreamless sleep, lucid dreaming, increased immune functions, deep levels of hypnosis or hypnagogic states, deep dreamless sleep, trance, suspended animation, miracle healing, divine knowledge, rebirth, trauma recovery, at-one-ness, samadhi, near-death experience, coma, Cloud of Unknowing. Anger & irritability can often be overcome by entraining a waking state in the Delta range. Delta in the waking state works well for crystal clarity, deep levels of non-linear thought, yogic meditation, lucid dreams, enhanced inner awareness, inner peace, energetics, sexual energies, provided one can remain lucid and awake during the induction. The Delta range is also home to deep trance states, pituitary release of growth hormones, some elements of self-healing.

4-8 — Theta Range — total recall, penetration of reality membranes, psychic intuitive state, clear far-reaching mental imagery, creativity, planning, lucid parallel universe voyaging, Zen attention, subconscious imagery, deep meta-programming, reduction of mental fatigue, sex drive, inner peace, mystic revelations, transcending belief systems, coincidence control, mind-control, deep inner healing, shamanic exploration, inspiration, deep states of high-indifference or "reverie", teleportation, jaunting, telekinesis, telepathy, creativity, inspiration, angelic or divine guidance. Chanting, drumming, spirit dance, short-term deep relaxation and restoration and shamanic voyaging occur in this range. Delta is far below the violent and suppressive activity of the Beta-Brain Functions.

7.83 — Schumann Resonance — powerful ESP and paranormal energetics, the prime shamanic psychic-intuitive state, lucid dreaming of the highest order, shamanic voyaging, rebirth and prior life memory access, penetration of parallel universe membranes, out of body voyaging, waking state, high universal resonance, one-ness, goddess resonance, angelic resonance, creative flow, divine inspiration, calm relaxation, stress reduction, heightened sexual experience, positive emotions, intense concentration, heightened awareness, dowsing, telepathic communication, angelic guidance, healing energies, dream interpretation, higher body and body of habits entrainment, between-lives training. Einstein is said to have had a sudden insight of the Unified Field Theory in this state, as well as many other great breakthroughs by Newton, Pythagoras and others. We owe most of our creative insights to the Theta State. Theta is generally totally suppressed by Beta.

8-13 — Alpha range — Dreamy but aware state of pleasant deep relaxation, useful for those interested in personal comfort zones. Highly prized by those very involved with the organic body. A floating daydreaming state of almost totally relaxed non-focused awareness coupled with a light, euphoric absorption in nothing in particular. Alpha is ferociously suppressed by Beta.

14-40 — Beta Range — Normal Brain State, evaluating data received through the senses, continual associative/comparative functions, with fearful conditions of worry, anger, fight-or-flight reactions, and a variety of organic, mental and lower-emotional hungers punctuated by periods of momentary satiation, further stimulated by shock and surprise reactions. In the Beta Range, all motivation stems from outer world expectations and fears, survival, problems, automatic sexual arousal and negative emotions. Dwelling on past injustices and wrongs, focus on avoiding mistakes, errors, accidents and personal humiliation. Beta is a high-frequency thought pattern of associative-comparative calculations which are in general logical, analytical, intellectual and excessively verbal. Beta is task-oriented and outwardly-directed, highly environmentally dependent and at the effect-end of the cause-effect cycle. Stress, anxiety and social dynamics are of the highest priority. Celebrity-fixations, naming of objects, bottom-lining and conclusions are dominant in Beta, as are sensuality and sensory stimulation. Problem-solving in fearful or dangerous situations can cause Beta to rise into the High Beta, 31-40 Range, in which intricacy, detail and integrative mental functions are demanded, causing the effect of a Human BioComputer, basically focusing on extreme minutia while sweeping on to the Grand Fallacy.

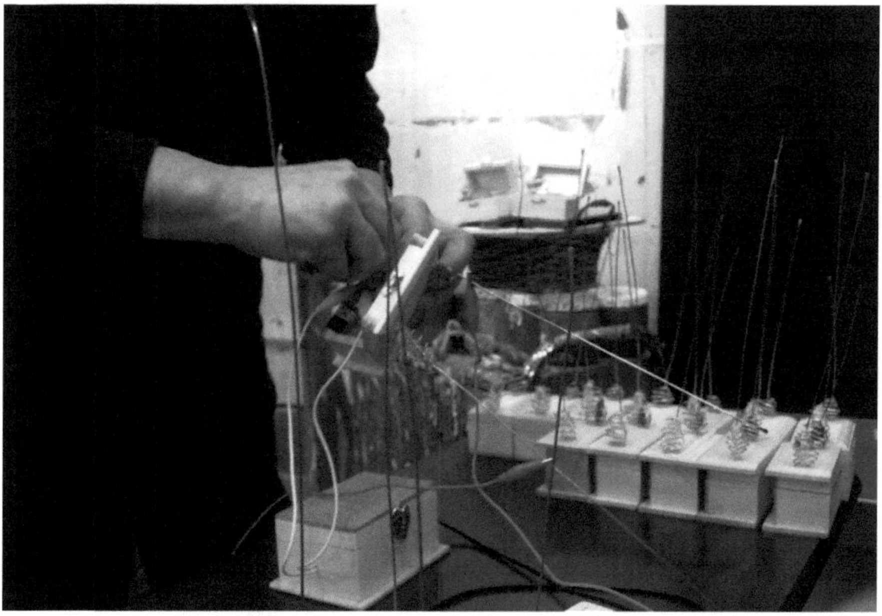

A variety of "tunings" can yield quite different results.

You can wear more than one Beta-Blocker at a time . . .

Beta-Blocker amulets look and feel great!

Chapter Sixteen

WEARABLE BETA-BLOCKERS

Several folks asked during last weekend's workshop if I could possibly come up with a miniature wearable Beta-Blocker. After four days of research on coil windings, capacitor and resistor values and other such sundry, I did in fact manage to make one, about dollar-size, that fits well into a sterling, gold-filled or 14k solid gold locket.

Then this morning, I came across a great site that has a radio coil calculator, and with its help, developed a 7mm airform coil that gave me the values I needed to end up with a waveform with an RF Resonance Factor of exactly 7.8 Hz. at the Schumann Resonance of High Theta, precisely where I wanted to land.

The germanium diode just barely fits in the watch-crystals, and I had to switch from acrylic crystals to hand ground and polished mineral glass watch-crystals to get the signal out in a decent micro-henry (uh) range, but it did eventually work the way I'd hoped it would.

I have allowed my demo of the apparatus to be made into a video for YouTube, and will demonstrate it live at our new Brane-Power center in Grass Valley, CA.

I think you'll like this item; it's not just a great piece of work tool, it really looks totally cool as a piece of jewelry, and attracts a lot of attention.

Wearable Beta-Blocker amulets bring out the inner beauty.

Chapter Seventeen

BLUDGEONS & SLEDGEHAMMERS

THE BLUDGEON:

If you've already gotten and started using your Beta-Blocker, you'll have noted that the Field Effect can be overcome by someone with a serious Beta problem.

Many folks are frightened of the non-Beta state; they don't know how to operate without the fear-driven calculations, associations and comparative functions of the Beta-Brain, and so they clamp right down.

The Beta-Blocker works well, but you have to LET IT DO ITS WORK. If you insist on holding onto the Beta-Brain, you can override the Beta-Blocker and continue right along in freakout mode.

I came up with a stronger dose of Beta-Block, which is a low-grade 100 milliwatt transmitter which broadcasts live subtractive resonance frequencies. In fact, unlike the Beta-Blocker, which has an effective range of about 4 meters, the Bludgeon has an effective broadcast range of about 1/4 mile. You could literally hook up a microphone or CD player and broadcast legally to your entire neighborhood!

This broadcast unit does NOT require an FCC license to operate. It is well under the .5 Watt limit, above which you DO need an FCC license.

The Bludgeon costs about $750 to build, and that's what you'd pay. I do not charge for my labor. The difficulty level of this broadcast unit is so high that it's beyond many highly skilled radio hobbyists to build.

THE SLEDGEHAMMER:

The Sledgehammer is for those blockheads who just can't let go of the organic game at all; who just for some reason seem to hold on grimly to the Beta-Brain regardless of provocation. If you're one of those Beta-Brain fanatics, the SledgeHammer is just the evolutionary tool for you!

The SledgeHammer is an actual professional-level broadcast unit that is capable of sending out music and voice over a radius of well over 2 miles, meaning that its signal is four miles across!

The composite reduction waveform hits literally like a sledgehammer. I know it sounds dangerous, but it is far less dangerous than walking around the streets with dozens of 50,000 Watt stations jamming your brain at all hours of the day and night, as well as the Russians with their damn "Woodpecker" and the Chinese with their psychojamming devices aimed at the U.S. and Europe.

If you need the SledgeHammer, it'll cost you about $1500. It takes about six weeks to custom-order to the correct subtractive frequencies to resonate at the desired 7.8 Hz composite-waveform.

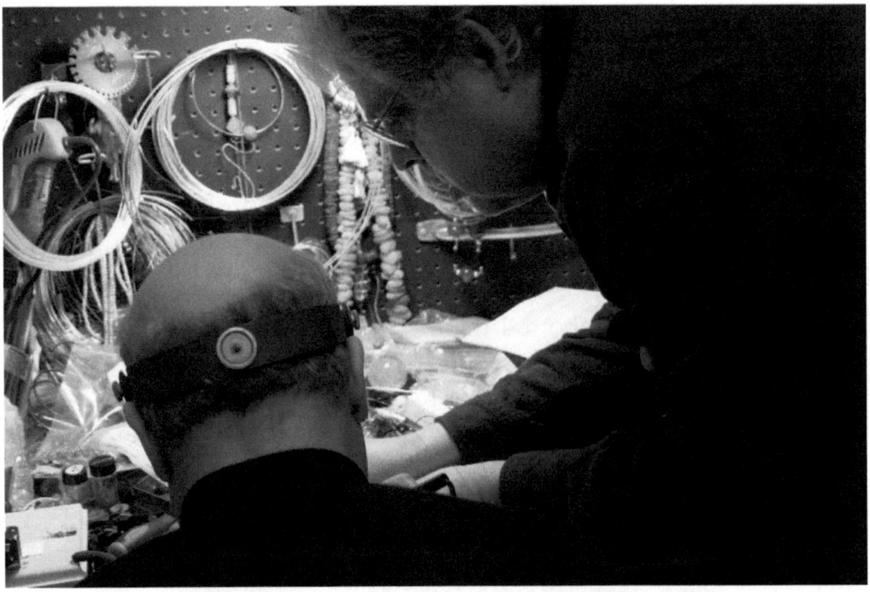

Claude and E.J. work on a SledgeHammer, 2009.

Chapter Eighteen

BETA-BLOCKER SEX

I thought that title might grab you by the short hairs. Actually if you think back on it, you'll realize that virtually every tool I've given you over the past fifty years — that's how long the Institute has been around, although it didn't incorporate until 1971 — is some sort of Beta-Blocker or depends on one or another type of Beta-Blocker to be effective.

Back in 1966 I started giving the Movements Exercises, along with such inner exercises as "A buzz buzz, 26 buzz buzz, B buzz buzz, 25 buzz buzz, C buzz buzz, 24 buzz buzz, D buzz buzz, 23 buzz buzz," and so on, up the alphabet and down the numbers, then reversing, throughout the dance movement exercise called Dervish Exercise #3.

Other Beta-Blockers have been:

The Black Box, the Big 24, Bardo Safaris, Street and Craft Fair Entrepreneur Work, Bunraku Puppetry, Comedy Workshops, Bingo Calling, Erotic Spirit Dance, Music CDs such as Ritual of the Cave, Inside-Outside, Secret of the Meteorite, and many more, Live TV Broadcasting, Inner Circle Online Workshops, Sacred Movements, Altar Work, Silver and Coppersmithing, Ceramic Sculpture, Watercolors, Pastel Work, Charcoal Drawing, Graphite Landscapes, Roma Plastilina Sculpture, Acrylic Painting, Zen

Gardening, Haute Cuisine Cookery, Fashion Design, Interior Design, Bluelining, Bonsai, Organic Food Gardening, Breadmaking, Gold-Dredging, Archaeology, Day-Trading Stocks, Sailing, Trail Hiking, Campfire Cooking, Barn-Raising, High-Speed Book Production, Electronic Circuitry, Zen Photography, Essential Ritual, Miniatures, Dollhouse Building and Furnishing, Woodworking, Woodburning, Leather Crafting, Model Railroading, Solar Power Production, Wind Power Production, Tesla Coil Making, Amateur Radio Electronics, Target Shooting, Compound Bow Making and Advanced Archery Techniques, Fabric Painting, Scraffito Ceramics, Ceramic Painting, Plush Toy Collecting, Advanced eBay Sales Techniques, Advertising and Marketing Methodology, Candle-Making, Incense Formularies, Essential Oils (aptly named), Bath Salt Mixes, Rare Book Dealing, Coin Collecting, Antiquity Dealing, Paper-Making, Fine Art Scrapbooking, Theater, such as *Creation Story*, *Redfin* and all the radio plays and readings we've done . . . and much, much more over the years, including Cybele's and my book for Simon & Schuster, *Beyond Sex*, which is the concept of Beta-Block using sexual energy, hence the titillating title of this little dissertation.

Sex has always been a great Beta-Blocker, but by itself it does nothing. If you expect more of it, you have to put more into it, and that's one of the items on this year's Workshop agenda.

We'll also be working with electronic Beta-Blockers, notably in the broadcast bands. As you may have noted over the past few communications, I've been busy developing several forms of the Beta-Blocker including a personal wearable device suggested by Denise Wey, and the Ghost Hunter Spirit Radio, the Cleanser-Charger for Body Workers and Hand-Healers, and two fun little items for hard-head Beta-Brains, the SledgeHammer and the Bludgeon, which are high-energy Broadcast Beacon Professional Transmitters in the AM Broadcast Band.

All of the tools I've introduced over the past five decades are intended as Beta-Blockers as well as learning tools to exercise your skills at split-second decision-making, emotional balance, team-work and, well, all the work benefits that you contributed in your chapter in my *Spiritual Gaming* book.

The ideal tools have finally reached fruition; the Portable Beta-Blocker goes on top of your bass-reflex speaker to the left of your computer screen and works as you go through your Bardo Safari and as you participate actively in our two-way online television Inner Circle Workshops and Quarterly Workshops, one of which is our traditional Easter Workshop which this year is all about Beta-Blockers.

You'll soon discover that Beta-Blocking is a powerful goal for us, and here's why:

The Alpha-Theta and Delta States are the natural states of the Essential Self. The Beta-Mind is imposed by organic life and the needs of the surrounding culture.

Beta-Mind does not come into play until the age of five — coincidentally the time most kids begin school. Schooling in the West is in general designed for one purpose and one purpose only; to prepare the child for the work-force, period.

The Beta-Mind is jammed on your head like an inverted aluminum colander, and it's about as useful in holding water as a colander upside down or rightside-up.

The Beta-Mind teaches you to fear the future, avoid the present and regret the past. It is conditioning, pure and simple, to program you to get up out of bed at a time you'd prefer to remain for a while longer, eat cardboard in homogenized milk, drink processed juice, gulp down scalding hot coffee and grab a bear-claw or a nice oily donut on the way out the door to a crowded public transport or to your car on a crowded gas-clouded freeway to crawl slowly in traffic to a place you don't want to go to, so you can work at something you'd rather not be doing, with people you really don't want to be with, for a company you'd rather not be working for, with a boss under whose thumb you'd prefer not to be, for crap money that goes so fast you need a second job just to make ends meet, all so you can unwind on the weekend so you'll be ready on Monday morning to start the whole cycle over again.

If you examine your schooling, you'll see that it did the job well. And your High School and College education were, if you'll take a moment to examine that carefully, designed not so much to teach you anything as to weed out the Alpha-Theta Intuitionals, the

Unwanted Facial Hair of the Work Force.

All that training and conditioning over so many years will take a bit of work to uncondition and retrain yourself away from. This process is called “Entrainment” and is what we'll be dealing with almost exclusively as we work toward penetration of the Phenomenal Veil.

Well, enough chit-chat, I'm off to wire up some Beta-Blockers to get them ready in time for our workshop. Hope to see you there; I think it's going to be something really special and useful for your work.

Love amulet can be used in tantric rituals.

Chapter Nineteen

EARLY LEAKY WAVE TECHNOLOGY

The Spirit Radio Beta-Blocker utilizes a full-wave integral equation which takes into account both reduction waves and periodic resonance along the EMS.

Laterally-shielded rectangular dielectric waveguides periodically loaded with planar perturbations of rectangular shape is the direct result of my early research in physics on open periodic wavetables.

The propagation of leaky-waves from leaky-wave closed-loop antennas can be used to cause resonant lower-frequency leaky-waves within the brain.

A new form of periodic leaky-wave antenna was used in my Beta-Blocker; its radiation patterns are triodic and can be analyzed in three-dimensional trinary trigonometric patterning using radii stemming from the hotpoint of the antenna radiating outward in the usual Stettman Dianometric Pattern similar to that of a commercial AM radio broadcast station's vertical mast, but the loop-back system introverts the waveform to a local area roughly four meters in radius rather than several miles.

The intensity of the resultant reductive waveform is diminished on the lower end of the EMS at 7.8 Hz, resonant with 1440 Hz on the broadcast band, through Beat-Frequency Oscillation and standard wide-wave technology common in radiophone equipment dating to 1902-1922.

This radionic technology was subsequently buried and lost in the rush toward miniaturization, transistors, integrated circuitry and more recently, binary cellular technology and satellite download systems.

Tube technology carries the older technology forward just enough to make it possible to utilize it for Beta-Blocking. The legendary Armstrong 1-tube broadcast band radio is ideal for a large-scale Beta-Blocker which could cover an entire workspace of about 10,000 square feet if not blocked by steel and concrete forms.

A double transmitter system can be built with an internal/external leaky-wave antenna radio-wave transmission system for industrial application in the workspace and can be tuned for Alpha Calmness and/or Theta Creative Inspiration in the workforce.

The double transmitter would send a compound/complex signal at 1440 Hz, and another at 1447.8 Hz, producing a compound reduction wave at 7.8 Hz, the target frequency for Schumann Earth-Plane Resonance. The key to this system is the antenna design and unique coupling of the two transmitters utilizing a mechanical Corpus Callosum imitating the function of the left-lobe right-lobe "highway" of myelinated fibers.

When memory is built using the Beta-Mind, memories and comparative data are in fact time-stamped when entering the hippocampus, while the prefrontal cortex sorts and files new information using the entorhinal cortical functions for long-term memory storage.

It is a demonstrable laboratory factoid that musicians and artists use both lobes more and more often than other people and that making music improves language skills without direct imposed learning. This is one use of the Beta-Blocker and may be part of the reason that focus and emotional stability are affected by Beta-Brain activity.

Active waking use of the cerebellum, meaning to remain awake during Delta Region Activity can improve learning and opens long-term and short-term memory "lanes". Sleep is mentally as active as the ordinary daytime waking state, and when the brain remains awake during Delta, amazing results can be obtained, so long as rest periods with sleep are not ignored.

Theta profoundly affects Broca's Region, and music and dance are clearly indicated as adjunctive activities for this activation. Again, the influence of the Beta-Blocker at the level of professional broadcast transmission, i.e., the Bludgeon and the SledgeHammer, can be an effective way of introducing the Waking State in the common workplace, using my special design leaky-wave low-band transmission antenna as a convection medium.

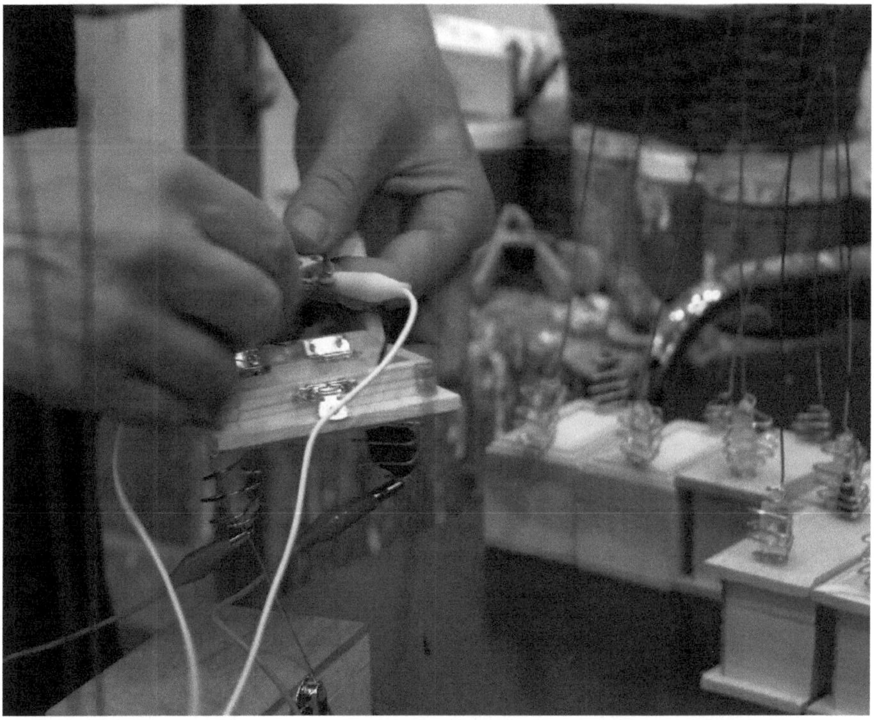

Leaky-wave antennas are a feature of tabletop Beta-Blockers.

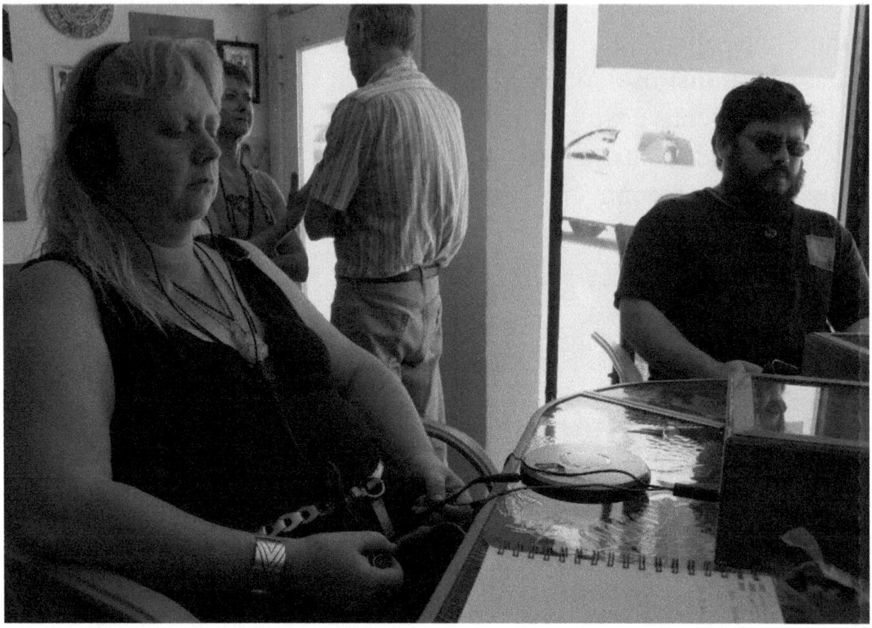

Beacon training at the Neal St. Center, 2009.

Beacon training using Parallel World Induction.

Chapter Twenty

INTRODUCTION TO THE BEACON

The Beacon is a highly modified crystal radio connected to a pair of copper-wrapped quartz crystals. It is not, nor was it ever intended for, the prevention, diagnosis or treatment of any disease whether physical, emotional or mental.

The Beacon consists of a glass-topped wood box, with a 1/4" phono jack in the front. Within the box is a hacked combination of Morgan, Dunwoody, Litz and Bucher crystal radio technology, terminating in a pair of psychically charged crystals.

The type of crystals in the grippers will vary depending upon the psychic-spiritual level of application, including native quartz, black tourmaline, fuchsite, amethyst and chondrite meteorite cores.

The crystal grippers are attached to the connecting cord with covered alligator clips. Should they become disconnected, simply clip them and pull the plastic covers down over the clip for comfort and aesthetic.

The crystal grippers are held one in each hand. It does not matter which cord is left and which is right — plus and minus orientation is not important unless you are training on higher levels, in which case the black cord goes to the left hand and the red cord goes to the right hand for meridian balance.

Because the crystal radio circuit derives all its power from the radio waves themselves, there is virtually no current running through the body except what normally comes through broadcast radio waves in Earth's atmosphere.

Broadcast radio waves in the AM band have several sources: man-made "intentional" waves, Earth's magnetic field and stellar radio waves including those from the closest star, the sun.

Magnetism and electricity are co-related forces, hence the measurable and detectible field which they generate is called the electromagnetic spectrum.

The Beacon is primarily a focus point for psychic-intuitive energies, a crossroad through which very subtle energies pass and are brought into relationship with one another through a coupling factor of sensitivity and selectivity. The Germanium crystal is called a "detector", but what it actually does is separate the AC — Alternating Current — of the radio waves into DC — Direct Current, which can then be processed by the remainder of the radio circuit.

You may feel a bit of a tingle when holding the grippers, not from any measurable voltage, but from the crystal energies themselves, enhanced by the copper wrappings and their connection to the radio waves processed within the Beacon's crystal technology processor.

Place the crystal grippers on the table in front of you. All lower level procedures are intended in the seated position, but if you prefer, you can lie down and close your eyes if you are using a Beacon at home or in higher level procedures.

Put the Sony headphones on — you will note that they are oriented with an "L" for the left ear and "R" for the right ear. You may prefer to use ear buds, or another high quality headphone, which is just fine. Everyone is required to bring and use their own headphone unit at a training center for the sake of personal hygiene and also for psychic-spiritual cleanliness.

The induction disk should now be placed in the Walkman CD player. Do not use an MP3 or copied disk for this purpose. The disk is itself "charged" shamanically and is an important part of the functionality of the Beacon.

Later more advanced training will not use CD induction. Guidance in higher levels is provided by the Certified Brane-Power Trainer in accordance with approved training methodology.

Relax as much as you are easily able to do at this time and drop the facial mask by relaxing the facial muscles. Release any tension in the shoulders and breathe normally, naturally and gently to get the best results; however if you are feeling tense, it's all right — the tension will be gone by the end of the induction.

Start the disk by pressing the "play" button to the right of the little window on the side of the Walkman. Adjust the volume as necessary with the "plus" and "minus" buttons to the left of the little LCD window.

Pick up the crystal grippers and allow the induction to do all the work. Don't worry if the Chronic fights the procedure — it works anyway.

When the induction is complete, which at lower levels will be in about ten minutes and in advanced procedures will be up to an hour, you can place the grippers back on the table and remove the headphones or ear buds.

You can continue to hold the grippers for simple Beta-Block if you like. You should never do more than one Parallel Universe Persona Training per day, but you can use the Beacon as often as you like for as many hours as you like for other purposes from Beta-Block to Alpha-Theta Inductions which you can use at the Brane-Power Training Center or purchase for home use at the center's bookstore or from the Brane-Power.com website.

When you have completed the training, you need to write down your observations in the Beta-Block LogBook. We request that you type it out for us and send the report by email so we can publish it using only your initials for the benefit of others, but this is not a requirement, just a request.

Before the next level of training, a Brane-Power Trainer can use a "Checker" device to determine whether any of the connections were missed, incompletely downloaded or failed to integrate fully.

Advanced levels will use the Beacon slightly differently. A Trainer will give instructions at that time on how to use the Beacon for Advanced Levels.

IMPORTANT NOTE:

There may be some very pronounced psychic-spiritual effects immediately following the training, including a feeling of being "high" or intoxicated. If so, do not attempt to drive or operate heavy machinery until this effect has passed. It usually lasts only a few minutes but can continue for longer in some cases where the trainee is highly spiritually sensitized and has had shamanic training or psychic-medium experiences.

The actual effect of the training may not reveal itself until later, when a skill or ability presents itself which was not learned in this lifetime track but was learned by a different Parallel Universe Persona on a different life-track.

The immediate effects will vary from individual to individual, and from one training level to the next, but the cumulative effect will be to "cluster" many Parallel Universe Personas into one highly charged and superpowered personal manifestation of a single unified higher self, called the Essential Self, the spiritual director of your personal manifestation in the HomeWorld Universe.

If you have any questions about the Beacon, you can email me through the Brane-Power.com website or enter into any of my chatrooms and forums. I'll be happy to answer as best I can. For pure theory, you can read my book on Brane-Power technology which you will find on the front page of Brane-Power.com, but please remember that in the field of physics, everything is speculation — even what has been "proven" by technical math "proofs" and lab results will change as we learn more.

It has been said that today's magic is tomorrow's science and that today's science is yesterday's magic. My Beacon and Beta-Block spirit guides are by those standards, 90% magic and 10% science.

I think you will be pleased with the results — fully energized and present with full attention, ready for life and all its surprises, ready for passage into other dimensions with eyes wide open and a childlike sense of thrill and wonder.

Chapter Twenty-one

BRAINWAVE BEATS

I suppose I should explain a bit about exactly how brainwave devices work in general.

If you've got a roomful of, let's say, grandfather clocks, and all the pendulums are swinging at varying points, eventually all the pendulums will swing in unison.

If you've got thirty women living together closely in one compound, as they often do in India and Western Asia, they will all be on the same menstrual cycle within a day or two of each other, and in some cases all on the same days exactly if they've been together long enough. I know you've seen this phenomenon.

Watch people's feet as they sit listening to swing music, and you'll note the phenomenon there as well.

Rapid light pulses such as strobe effects at a Rave will cause the brain and nervous system to resonate with the light pulses.

Heartbeats will eventually fall into rhythmic patterns together. Romantic, ain't it?

And of course we all know what happens when you strike one tuning fork and hold it next to another, but did you know that the same thing happens on a quantum level?

You can see this phenomenon of pulsing with a driving persistent rhythm in drumming circles. Eventually everyone starts dancing or moving at least a little to the drumbeats.

It's the same with brainwave "beats". You set up a pulse at the target frequency or a resonant factor of your target frequency, and after a while, everything begins beating at the same frequency.

In addition to that, add a suppressive signal to the Beta range and you have what we call the Beta-Blocker.

It's a gentle rhythmic ebb and flow, a slow swelling pulsing beat at the right frequency to get the Beta to relax into submission.

The whole family can gather around a Beta-Blocker and play Scrabble, Perquacky or Monopoly and use the time spent in that play space to entrain the brainwaves into Alpha-Theta Range.

How about a party with a Beta-Blocker as the Guest of Honor, in the center of the room on a pedestal???

By the way, it occurred to me as we get phone calls from folks saying their Beta-Blocker has arrived in Europe and Canada, that these things were actually transmitting as they went through the postal service. Maybe "going postal" will be a thing of the past if we can get enough of them mailed out.

I'm currently working on a Master Blaster Beta-Blocker, which can be operated as an unlicensed transmitter under FCC Part 15 rules.

Well, I'm off to dinner and some rest. I've been crafting up Beacons all day and part of last night, and I'll be doing the same at about 1:00 a.m. tomorrow in anticipation of the large gang that's arriving in the afternoon and evening.

I'm delighted with the stories some folks have told about their first experience with the Beacon, and I'd love to hear yours!

Chapter Twenty-two

PARALLEL UNIVERSE PROBES

In Parallel Universe Probing using the Spirit Radio, we can easily and instantly "know" the distance of the target universe by exact number of Kakus.

We can also identify the exact amplitude of signal relative to all other signals from all other universes within our scope or "reach".

How do we do this?

The Beta Mind cannot know these things. It depends entirely upon associative data and similarity-constructs to "know" the little that it does know.

The Alpha Mind is too disconnected to "know" anything. It just digs where it is and has no focus whatsoever.

The Theta Mind is what "knows" the distance in Kakus, "knows" the amplitude of signal and "knows" the exact thoughts, emotions, sensations and environmental images connected with extra-universal events in other Parallel Worlds as well as our home dimension of the moment.

Don't forget, we travel instantly in quantum "jaunt" effects from one parallel universe to another all the time. Why don't we know that? We do, as children, recognize the space changes. With the development of the Beta-Brain, we lose contact with the abruptly switching realities.

There is no process of "knowing". The Theta Mind is and always has been capable of knowing these things. It does not need to be trained to "know", but in order for us to trust our Theta Mind knowingness, we must entrain ourselves to TRUST the Theta Mind's knowingness, which is somewhat tough for those saddled with a hard-boiled Beta Mind that digs itself into a hole of cynicism, sarcasm, suspicion, fear and worry.

So how do we know these things? We just do, that's all. How do we learn to trust this knowingness? That takes some skill and a lot of reinforcement. We must many times over ACTUALLY SEE that our Theta Knowingness is real and accurate, and that takes time and lots of experimentation.

The experiments related to proofs that the Theta Mind does indeed have the right answers is called "evidence gathering" and is mostly what you'll be doing in the early stages of your Theta Mind training.

I hope this helps those Beta-Tortured folks who asked, "How can Oz possibly know the answers to those questions???"

Parallel Universe probes at Neal St. Center yield powerful results!

Chapter Twenty-three

PUP TRAINING — GROUP EFFECT

You'll notice that a group of individuals gathered in a PUP training space will cognate on the same events, distances, amplitudes, numbers and descriptions all at the same time.

What this means to you is that you will know the number of Kakus from the Home Dimension before the subject in an experiment says the number he or she has cognated on.

You'll have the answers actually faster than the speed of light, because Theta Mind cognitions occur instantaneously by Spooky Action rather than at the speed of light, as would energy waves or wavicles.

The net effect of this is that if ten people are gathered around a PUP training session, and they independently note down the numbers, provided they are all "tuned in" and not impinged by Beta, the numbers will all be the same.

What happens in the Beta-Brain is that the "first hit" or initial spontaneous knowingness is rejected, ignored or bypassed, and the automatic "I don't and can't know that" of the Beta-Brain is substituted for the knowingness of the Theta Mind.

This same effect will, as many have noticed, obtain over an internet television transmission as well as it does in person.

I am working toward developing a training program so you can do this training with others in your own Home Town. The training program will be set up in stages — Level 1, Level 2, and so on, with specific depth of penetration and extra-universal contact and strengths of probe contacts.

DO NOT ATTEMPT THE PARALLEL UNIVERSE PROBE without first learning how to apply the Remedies. When you mix two or more quite different realities and swirl it around in the brain, you can get pretty far out there, and sometimes that's not entirely manageable. The REMEDIES are very specific and relate to Life Guard Duty in terms of application.

Parallel Worlds explored by Tequiha Foundation shamans, 2009.

Chapter Twenty-four

PUP TRACKS

Every crossroad in life leads to at least two possible world variants. The branching process can become quite elaborate and some of the branches are quite dangerous, such as those which lead through the variations of influenza upswings.

In brief, the universe is filled with bacteria and viruses, which are in continual warfare with one another. Any temporary passage through the zero point is only momentary. There is no "steady state" at the zero point, yet $e=mc^2$, which is to say, the equation always equals zero with balance trade-offs on both sides of the equal sign.

In the case of viral/bacterial warfare, it is important to note that both viruses and bacteria are capable of nearly infinite adaptation, even to zero gravity and zero atmosphere.

Viruses are not exactly alive — they are a form of "half-life". Bacteria, which are the basis of human and other organic life, are by their own definition, and only their own definition, "alive", in the sense that they answer to certain definite and narrow criteria which taken together give meaning and significance to the term "life".

Organic life is always subject to viral attack. Bacteria attack occurs within the spectrum of "life" and is therefore subject to "death", meaning that bacteria can be killed. Viruses cannot; they suffocate into submission, they are subject to remission and can starve themselves away from a dominant strain, yet they are never quite "dead", just as they are never quite "alive".

Viruses can be thought of as a form of self-replicating nanomachinery, capable of traveling through stellar formation and dissolution, and of voyaging through vast reaches of interstellar space over unthinkable lengths of time and curvatures of local magnetized special fields. What this means to you is that you are traveling in a constant war-zone, no matter what "lineage" of dimension within which you might find yourself at the moment.

Each line of reality has its own place in the war zone and each has its own brand of susceptibility to attack and resonant vulnerability to viral invasion. Note that uniquely, viruses kill their hosts — bacteria tend to enslave them and keep them alive as food sources, much like slime molds, which "farm" their food sources just as humans do.

When you find yourself in a bad time frame along a particularly dangerous route in the reality spectrum, it might be wise to "change tracks". This can be easily accomplished through the use of the BEACON AUTOMAPPER.

The BEACON AUTOMAPPER seeks the hottest "next-of-kin" parallel universe persona (PUP) and can either bring it in or traverse you there; the survival mechanism within your organism directs this process so that you will automatically seek the safest route through alternate universes.

Should you find yourself in a particularly dangerous alternate universe, such as that which hosts the deadlier strains of SARS, one simply sets up the BEACON and hooks up to the parallel self in the next-nearest alternate universe that does not contain the threat.

Remember that the BEACON will automatically MAP to the safer universe — not the absolute safest, but safer than the one you're in.

The BEACON uses your electrolytic imprint in the contained saline solution plasma you call your "body" to find its next-nearest neighboring persona in whatever universe in which it currently exists.

This effectively causes you, the Voyager, to "change track", to switch presence from one alternate universe to another. You're already jaunting from one alternative reality to another all day, every day, so why not make it a safer journey by finding a safer landing pad?

PUP tracking, Neal St. Center, 2009.

PUPs are integrated over a 24-hour period.

Koyote and friends in Soul-Retrieval Workshop; Neal St. Center, 2009.

Soul-Retrieval Workshop, Neal St. Center 2009.

Chapter Twenty-five

THE ASSEMBLY

Some call it "The Assembly", some "Re-Unification" and some have no name for it at all, but whatever it's called, it means simply to re-unite your various "selves" that are spread all over a number of Parallel Worlds . . . I'll explain:

Whenever you feel a tug to go in one of two or more directions in your life, what is commonly called a "crossroads", there is a definite split of realities, of dimensions.

In one dimension, you decide to take the left path which leads to Dimension A, in another, you take the right-hand path which leads, let us say, to Dimension B, both of which already exist; you merely choose which one you are going to live in for the time being.

From that moment on, you exist in at least two definite and quite distinct "worlds" or dimensions, which are now commonly called "Parallel Worlds".

In each of these Parallel Worlds, your path will lead inevitably to further splits, which we will call "decision points", each of which further leads to alternate universes in a descending path called "branching".

By the time you've reached the age of 40 or 50, you have spread yourself like butter over toast into at least 28 to 54 definite alternate universes, and you've got a lot more of your bodies scattered throughout the mega-universe from past and future lives — time doesn't actually exist except as a measure of the curvature of space, but that's another story entirely.

Your energy and focused attention and presence are, through this scattering process, distributed throughout a number of Parallel Worlds, thus reducing greatly your ability to manifest in this or any other dimension.

If you were able to re-gather your various "lost" parts, you'd have a lot more attention, presence and will than you have now, however much or little that may be.

Just ranging around and trying to find these lost parts is like sailing on waves of uncertainty on an ocean of probability — meaning that you'd be looking for a couple of dozen specific needles in a vast, almost infinite field of universes separated not by space but by frequency and waveform variance, which has to do with leaky waves and beat frequency oscillation combined with harmonic resonance and wave-cancellation. But more about those ideas later.

In the meantime, what all this means to you is that if you could collect yourselves into one big bundle of self-ness, you'd be a lot more powerful and in command of your voyage than you are when you're spread out all over a number of isolated and disconnected Parallel Worlds.

For this job, you need something like a GPS, which is a combination of map and navigation tools to find out where you need to go and to trace out a route to get you there safely and efficiently with the least number of side-trips and wrong turns.

That's where Brane-Power Training comes in handy. We've got The Map and the navigation tools and all you have to do is apply them.

Within a matter of only a few hours, you'll be well on your way toward re-unification of the separated selves, and before you know it, you'll have the power of The Assembly, which is the completed essential self, the "Single I" manifested in many worlds over an infinite stretch of time and space.

Chapter Twenty-six

COMPARISONS BETWEEN BETA-BLOCKERS

I'm taking a few minutes away from my seemingly endless round of soldering and circuit-design to explain to those who wrote in questions about the Beta-Blockers . . .

First of all, the wearable Beta-Block Amulets are entirely personal. I've taken the extra capacitor out of the circuit — it can be variable or fixed to a specific frequency — in order that the natural capacitance of the skin and electrolytic field-effect of the human biological machine can absorb most of the milli-micro and nano-Henrys of the small coil antenna built into the amulet's crystal enclosure.

I did this to keep the device entirely personal — you really don't want to influence people around you without their consent. It's a karmic issue that I take very seriously.

The Black Diode in hand-wound silver, gold-filled and 14K gold bezel lockets are another issue altogether. They are incredibly powerful and should be used with extreme attention to their possible influence out to a range of about four meters, roughly 16 feet radius, which is sufficient to subdue a room full of partying morons.

That having been said, the field-effect of the Beta-Blocker Tabletop is very strong at about four feet, and extends outward in rapidly decreasing effect to a total distance of about four meters, roughly 16 feet.

The Bedside Beta-Blocker for Lucid Dreaming has about the same field-effect strength as the Tabletop model.

The Healer's Friend has an effective range of only a few inches between the leaky-wave coil antennas, but the effect on a hand passing through it is as feelable to a psychic-sensitive healer as if you'd scraped your hand with two pieces of 180 grit sandpaper.

The Deep Delta Dreamwalker has an effective range of about four meters, but is best used at a distance of one meter or less, which is a bit more than a yard away. The field-effect drops rapidly to the square of the distance away from source.

The Bludgeon — ah, the Bludgeon. You can set that puppy up in the middle of a training center and it will handle the whole thing at a range of the average Victorian 3 bedroom 2 bath three-story house, which makes it great for the typical household if you want it to work that way. It can also be set for Deep Delta by turning the rotary switch to 1 through 4 for the four ranges of Deep Delta Induction Voyaging and Dreamwalking, or set to 7 or 8 for your Theta psychic-intuitive, should you wish to use it in the creative studio or whilst programming some toughie C++ assignment for your local game development company.

The SledgeHammer is a very different tale of two cities. What I mean by that is that it could be set up midway between Grass Valley and Nevada City and with the proper base-loaded antenna mast at a bottom residence of 20 feet off-ground, it would profoundly affect every living creature and a few dead ones within an effective power range of about four miles across, and it has a higher power output setting that could reach a whole lot further, but for that, you'd need an FCC license. As it is, if left unmodified and kept strictly at my settings, it conforms to Part 15 of the FCC code for an unlicensed AM transmitter and is legal to operate without a license or permit.

Unlike the other devices, the SledgeHammer is high-maintenance — some components operate at high temperatures in this unit, and will definitely require periodic replacement and re-calibration every few months. Calibration is tricky with this unit because the "hot" operation differs greatly from the "cold-start" operation and adjustments can only be made under "hot op" conditions, making bench tests difficult if not impossible. It must be tested under load, fully installed in something, so I've made an outdoor transmitting test tower to calibrate the Sledges.

I am currently working on a variant of my Beta-Blocker devices which will be the strongest I've made, called "The Gemini" because it is a twin system unit as opposed to the double-usage coil system in the other devices, but the circuit is still giving me trouble. I'll keep you posted on my progress as I develop it out and work out the bugs. It won't go out until I'm satisfied that it will work effectively and efficiently.

I suppose I should explain how the double coil system works, but I really don't know. I've inherited the concept from a vintage 1908 crystal set that uses the same coil for ground loop as it does for the antenna tuning, and apparently both the antenna tuning and ground tuning operate independently of one another, while both are held within the same tight-wound 12-tap coil. I can even cross the wires and they seem to have no effect on the opposing tuning circuit, yet they're both attached to the same coil, albeit from two different ends with the tap somewhere in the middle. I suppose they're independent because their frequency focuses are so radically different that they are like switching left and right hand parts on a piano, but actually I'm just guessing at why they work. The net result is that you get two radios in one coil.

I'll be in studio this morning with Oz making the Level 6 Parallel Universe Induction, as well as working with Viko on http://www.yoyodyneindustries.com to make it better and more complete.

Beta-Blocker induction CD.

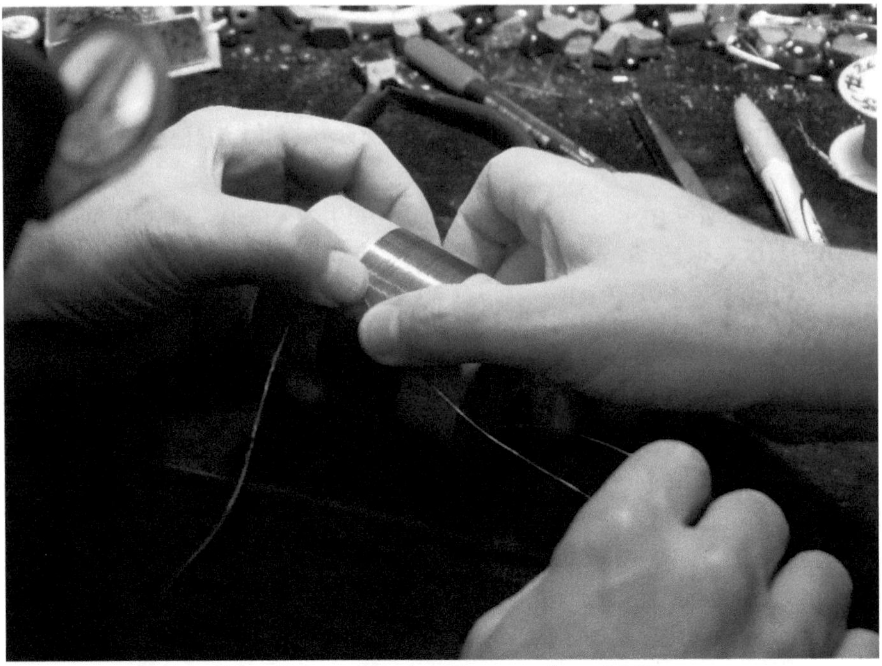

Coupling system coil wrapping, 2009.

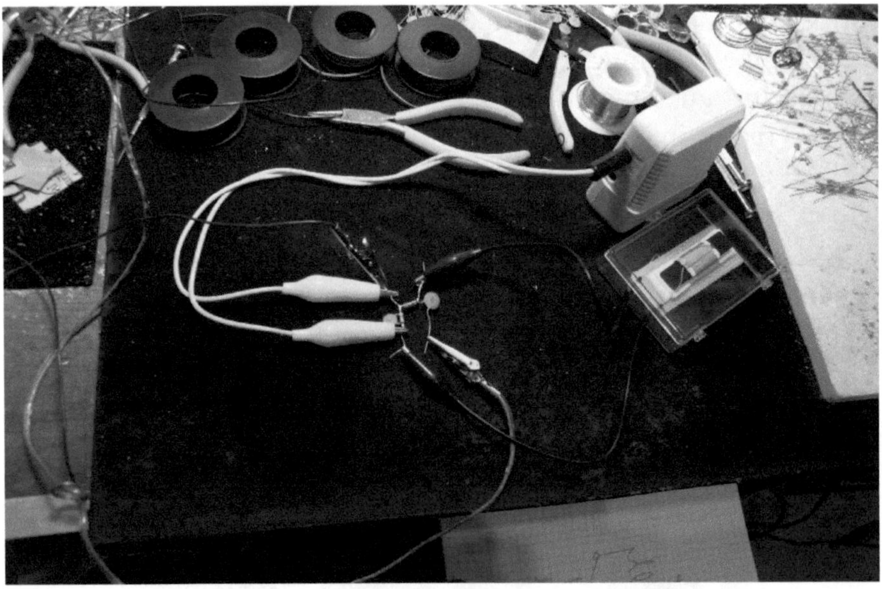

Coupling system on the test-bench, 2009.

Chapter Twenty-seven

COUPLING FACTOR THEORY

Optimization of negentropy of the Essential Self drives its attraction for neighboring PUPs.

There are bound to be random de-tunings of the electrical signature of the local time-space HBM which causes the Essential Self to desperately seek PUPs.

These random de-tunings can account for "waking state" and transference to alternate realities or tunings to alternate realities.

De-tuning causes tuning.

Tunneling and rat-holes (wormholes across dimensions without space-time continuum involvement is called a rat-hole) can account for non-linear skills and knowledge factors with no rational cause-effect in local personal history, meaning that skills and knowledge will suddenly appear within oneself with no visible means of support.

The blueprint of the Essential Self is very much like a virus in the sense that it will automatically and continually seek a host and alter the host to match its electronic signature, through a coupling factor, in which two fields will de-tune to produce a third electromagnetic field that detects both the "feed" or source field and the "eating" or sink field, although both anode and cathode fields are invisible to one another without the third intermediary field as a coupling factor.

Self-Replication of the Essential Self in automatic resonance field-effects produces overlap projection of co-influence through a phase factor that is the harmonization of mutually orthogonal non-projected waveforms, by the process of selective susceptibility, depending upon a common vector or carrier medium.

Electromagnetic co-absorption of mutually non-detected co-tuned waveforms depends upon QRM suppression of signal for wiggle effects.

PUPs are not seeing one another, but the Essential Self opportunistically uses automatic leaky-wave de-tuning of the Human Biological Machine to locate similar imprint signature PUPs in relation to the HomeWorld waveform's signature — in short, the advantaging of negentropy or syntropy in accordance with the Hellman-Feynman Theorem, producing a general effect of free enthalpy in the Planckian sense.

Catastrophe Theory provides two equal but opposite solutions that are dramatically different that can be flipped by a relatively minor influence.

Hot Bath Theory says that a heat source will organize itself into localized fields not in homogenous decaded order to the heat sink. These localized organizations are the most efficient route to redistribution of heat to a general field entropic release state.

Prototype-Circle of Peace amulet.

Chapter Twenty-eight

PSYCHIC INTUITIVENESS

The interesting thing to me about the Beta-Blocker is that it was a total accident, which is one of the best kinds, I think, of scientific breakthroughs. According to Willy Ley, our science editor at *Galaxy Science Fiction Magazine,* the great breakthroughs in science always occurred outside the laboratory — not the private laboratory, the corporate laboratory. This was Willy Ley's words: "It either takes place in a garage or the shed or the attic or the basement." And that's exactly what happened. I was in the barn at the time. I made seven prototypes of the Spirit Radio and number three prototype was the one I was happiest with, worked the best, gave the best reception, the best clarity and was the most sensitive and selective. In short, easiest to tune and had a good strong signal. As I was using the number three prototype, I noticed when I hit the 1550 and the 1507.8 frequencies that there was a definite Beta-Block effect with very strong high Theta.

Now, here's an interesting question. I got this question from a number of people which is: are the frequencies significantly different in Europe, or eastern Europe, or Russia, or Argentina? The answer is no.

When I worked at Columbia Records we were on Sunset Blvd. That meant there's a lot of street traffic, tons and tons of rumble, thousands of vehicles passing by every minute. In order to record, what they had to do was place a microphone outside the building which brought that rumble in electronically and then they used it to cancel itself. It's called wave cancellation.

ELFs are extremely low frequency, which means the wave passes through zero only a few times a second and the wavelength is 10 miles long or more. It's an enormous wave. A wave like that does not happen as frequently as a wave that is in the micromillimeters which is high-frequency, meaning it happens a lot of times per second. Extra low frequencies — ELFs — tend to travel through the ground, not the antenna. John and Toni Lilly discovered that whales are able to communicate using very low-frequency waves, and they can actually transmit a coherent message literally around the globe in 6 minutes using this low-frequency communication.

I'm looking at ELFs now because that's where I think we will see extra dimensional transmissions, more than in the high-frequency range. The high-frequency tends to be line of sight modulated by frequency, and low-frequency tends to travel through all kinds of things, and is modulated by amplitude.

Wave guides are little tubes that waves can travel in, similar to an envelope carrier at the right frequency which will carry that particular wave if it's tuned.

The music that was playing on my Mega-Wave Broadcaster was *I Can Free You,* which is a Beta-Blocker using audio signal. So between the audio signal and the radio frequency signals, you ought to get a hell of a good Beta-Block there. The Beta-Blocker doesn't use audio. A Beta-Blocker is putting out energy in the form of radio wave field effects. It's a field effect generator. So it's totally different from an audio signal which could serve as a Beta-Blocker to some degree, but is not nearly as effective as the radio wave effect.

For instance, if you were to sit in a live performance with Pablo Casals, you would find yourself in Alpha or Theta. There are some uplifting performers who can bring you to an Alpha state, which tend to just completely quell the Beta function to the degree that you're sitting in Alpha. Those typically are not the same performers who will leave you tearful and in a highly emotionalized negative state, such as Patsy Cline and Phoebe Snow. That's not a Beta-Blocker, that's a Beta Enhancer. But there are some audio-emotional Beta-Blocker performers — Pablo Casals is one, Yasha Heifetz another.

Bidu Sayao once in a while could put you in a really elevated state where your Beta was blocked to the extent that you were sitting in Alpha. That's not true of Victoria de Los Angeles who sang basically the same repertoire. There are some dancers who produced Beta-Block once in a while. Yma Sumac had that effect, and so did some performances by Led Zeppelin and the Who.

It kind of loses some of its sexy mystery to find out that all they're doing is suppressing Beta and thus transporting you into an elevated state. Keep in mind that Alpha-Theta is not something that isn't happening and then does happen. It's just that Beta is dominant in you after you've gone through grade school for one year. You've been entrained, totally programmed, to function entirely with the Beta-Brain, and you are reinforced constantly to use Beta, and to trust the Beta-Brain to guide you.

I produce Beta-Block all the time. I can transmit Beta over the airwaves, through music, through personal contact, and cause you to sit in Alpha and Theta wondering what the hell happened. Some folks have *always* lived in the Beta world, a world without magic and wonder — so when the Beta is gone, there is nothing left. They feel very lost.

Some folks have really well-developed Theta. They tend to be tarot readers or psychic mediums who have intuitional flashes all the time.

Claudio Naranjo is a psychic intuitive who does not have a lot of Beta development, never did have, although he was a very famous surgeon. He also is a very, very famous concert pianist in Europe and South America. He's not well-known in the states for that.

Heather Valencia is a psychic intuitive who has hardly any Beta development who functions very well in Alpha and Theta.

Dr. Laperla is very psychic intuitive. One of my art students Marci is very psychic intuitive, extremely so. So is Heather, very psychic intuitive. Maggie is very psychic intuitive also. Most of my students in the art classes are heavily psychic intuitive.

Here's what I see as the role of art in human society. If you have a very well-developed ability to look past the Beta veil: you can paint what you see, draw what you see, sculpt what you see, write what you see, poetize what you see, and you're penetrating the veil. You're always passing through this veil, you're seeing past it all the time. Just as a child does who has not been Beta-Brained, who's not wearing the governing cap of the Beta-Brain. If you are an artist, it seems to me that you would want to be able to penetrate the Beta veil to see beyond the consensus reality of human culture. Therefore, to be able to project into other dimensions, other realities, one would be actually external to the organic self as a natural state of existence, almost always out-of-body in the Parallel Worlds.

Now there are a number of folks around who live like that all the time and they do wonderfully well . . . *wonderfully* well. They are the creative individuals; there are some scientists who are in that condition. John Lilly was notably one of those who was a scientist who was also a psychic intuitive. Forry Ackerman was very much in the Alpha, not so much in the Theta, but very much in the Alpha. He was blissed out all the time and he was like a little child, like a 6½ foot tall little child. Vincent Price I found to be very psychic intuitive and very low Beta. He was a friend of Forry's whom I met quite a few times and had interesting chats with. He was very much non-beta. Mickey Dolenz is definitely an Alpha-Theta.

Most New Age scientists are super-Beta, producing all sorts of cool and groovy products that some Alpha-Thetas could possibly work with, but they have this buzz that is so ultra-cerebral.

There are several signals. Like the embryo who senses and reacts to ultrasonic waves. You can sense a variety of waves operating in people. Don't forget that the body is a large percentage of salt-water, and that water is constantly in vibration from internal wave producers. You can absolutely sense when someone is in Beta, when they're in Alpha, when they're in Theta. I can give you lots of examples, but I'll tell you one that everybody knows, which is called "contact high".

In the '60s someone would drop acid and walk into a party and everybody would go into an acid space. They'd all go into a contact high from one person having gone into that space. It's infectious, absolutely infectious.

Alpha-Theta is infectious and very high Alpha-Theta tends to suppress Beta around it to a distance of about 4 meters.

I'll throw out a Beta suppressive right now and you see what happens to you just sitting there. Notice that a lot of your "worry" apparatus just goes away suddenly and see what happens to your intuitive and psychic functions right now. I'll do this for a couple of seconds. Don't forget to breathe.

So as you see, you can be reached through air waves, video waves, television waves, and that is really a very important secret that I employ. You can leave your computer browser tuned to gorebaggtv all day long. No matter where you personally are, physically are, it will reach you, even if you're miles away at work. A very simple method which you can try at home is to simply enter into the Alpha-Theta state and hang around some people and just see what happens to them. They will suddenly lose interest in thinking and chatting and considering and comparing and associative minding, and just start kind of blissing out and seeing things a little differently. They're peering into an alternate universe or several of them at once, which they've been rejecting because the Beta-Brain's job is to *reject alternate realities*. That's what the "education" system is designed to do — to train you to accept the work ethic, the work force, the work space. You're supposed to actually become addicted to going to work at 9:00 o'clock in the morning, leave work at 5:00 pm. And to endure this until the weekend. And during the weekend you are supposed to be guzzling beer so you can suppress the Alpha-Theta and enjoy the football game in total Beta, with full memory of all the team and player statistics.

By the way, beer suppresses Alpha-Theta and also Beta, leaving you with a sort of semi-Delta, which is to say, almost unconscious.

Psychoactive is often associated with psychedelic substances, but psychedelic substances are not necessary to create a psychoactive state. Psychoactive simply means that you're in the Alpha-Theta state and that Beta has been successfully suppressed leaving you actually at that point functional; with psychedelic substances you are seldom functional.

The Alpha-Theta is infectious, transmittable, just like laughter. You start laughing, everybody around you is going to be laughing very soon. If somebody's yawning near you, you want to yawn. You're not tired, nothing is happening, nothing has been happening, but suddenly you see somebody or hear somebody yawn and you yawn also. Somebody starts laughing and you start laughing as well. And there are a number of individuals like Merv Griffin, who had Mrs. Miller in the audience to lead the laugh. Milton Berle's mother led the laughter in his audiences. And in some cases they'll use a laugh track on studio monitors to encourage the audience to let go and laugh.

In entrepreneurial enterprises you go out on the street with a booth of some kind, presumably selling something, but you don't really care what it is.

Part of your work is to infect people with Alpha-Theta by the process of Beta suppression. If you're a body worker you have an ideal opportunity to put a Beta-Blocker in your work space. If you are teaching Tai Chi put one of these in your Dojo. If you're working at IBM put it in your cubicle. Four cubicles away they'll still be affected by it.

ELFs go through everything, absolutely everything on the planet — in fact, they're detected by super long antennas that are buried deep underground.

It's kind of like measuring lightning. You measure lightning not by the strike, but by the upward lift. Lightning travels from the earth to the sky — it actually jumps upward. The ground is more important than the antenna. We drove an eight foot ground into our testing system.

We have a very long antenna that we use to calibrate and null our Beta-Blockers, Deep Deltas and of course our Big Guns — the Beacon and the Mega-Wave Broadcaster. Not only do we build Beta-Blockers, we actually test them, calibrate them and null them. I was going to mention frequencies because so many people have asked about this. Is the frequency going to be significantly different in Sweden than it is in the United States? The answer to that is no and there is a reason for this, which has to do with assignment of frequencies.

In the United States the FCC regulates the frequencies of various television, AM, FM, shortwave and microwave transmitters. But on a clear night, you can hear Radio Moscow. There is no other station near it that's going to interfere with it in most places, but in some places where there is a strong overwhelmingly dominant station, you can use the Bucher QRM Suppressor to null it and open the way for distant transmissions.

If all you can hear on your AM receiver is the religious channel, and it actually appears four or five different places on your AM dial because it's blasting out with so much power, which many of them do, you can null it with a QRM suppressor.

One of the questions you should ask yourself is when the Beta does go down, how do I feel and what do I intend to do? What tends to happen with me? What am I aware of? Am I functional at all or do I go completely non-functional and totally haywire with lack of Beta? It is a very important question for you to ask yourself and to answer.

If the answer is that you are disturbed or non-functional by the Alpha-Theta; if it feels uncomfortable to you or it feels like you're not functional in a highly sensitized psychic-intuitive state, you need training and you need the training *now* in the non-Beta state because in the voyaging between-lives you will have to function without the Beta — there is no Beta. The Beta is gone completely. Your mind, as you call it, is gone. It leaves you without the Beta to rely upon.

So you need to die before you die, to learn to live without Beta. That has a lot of meaning when you understand that what you need to do in order to prepare for the between-lives state is to get yourself into a state which is similar to the between-lives state now while you have a chance to. Of course nobody wants to handle an emergency until it actually happens, so the chances are that you'll wait until you die before you try to master this. Handling an emergency during an emergency is impossible.

If you look at certain other systems that are workable functional systems for learning Alpha-Theta, the only thing they teach you how to do is to sit in lotus with Alpha-Theta. The minute you get up from your lotus, the Alpha-Theta is gone and Beta comes slamming back in.

What I like to see is somebody who actually conducts business in

Alpha-Theta. Go to work, whatever job you have, and be in Alpha-Theta and be totally functional. If you can't do that, you need training. When I get into a Bardo Safari you'll notice that your Beta is remarkably suppressed, that you're in Alpha-Theta quite a bit. What I do is I project the Alpha-Theta through the medium of the game. When I do that, you can easily handle the stress of the game, but the average individual is going to end up going shopping.

I gave a number of examples years ago by projecting Alpha-Theta state and we would have a group of individuals, for instance, at a workshop, and some folks could handle it very well and they did fine. Others would sit there — you'd ask them a question and they'd say, “I don't know. I think I'd like some ice cream.”

When I had a thyroid storm, it threw the body into “fight or flight”, which causes extreme heart palpitations which can throw you into a heart attack. Some folks, when they get into the psychic intuitive Alpha-Theta state, will go directly into “fight or flight”. They'll start fighting around themselves, fighting for Beta-survival. Or they'll get violent and literally turn tail and run away.

Einstein said that the universe is a single unified field but that there are special smaller fields within it that don’t obey the general rules.

He used a special field theory to cover those alternate things that he couldn't cover with the unified field. Spooky action at a distance is based upon the idea that if you spin something levro in one part of the universe, you reverse the spin from levro to dextro on something which could just as easily be at the other end of the universe as close-by. Spooky action at a distance implies a single unified field. And in the waking state, you will definitely experience oneness of everything, the equality, Einstein’s Unified Field.

I live in a multiplicity of dimensions and I handle that fact very well as you can see, but that's through many kalpas of self-training and discipline to learn to handle my multidimensional existence.

I know you've had this experience. You've been driving for 20 hours and suddenly you're seeing double. What if you were seeing so many doubles that it was actually hundreds of different shifting patterns, all different. It would be like watching a movie projected on a movie, projected on a movie, projected on a movie, projected on a movie, projected on a movie, projected on a movie, all around you. Your reality structure, your understanding and personal history, are also multiple. In the end you'll have thousands or millions of personal histories packed into one single manifested Essential-Self, all of which are equally true. Many, many thousands of realities all of which are equally true right now, thousands of world perceptions all equally true, all of which are demanding equal amount of attention, equal awareness, all at the same time, and yet you've got to pick up your corn puffs at the supermarket. When you go to the store what do you buy for yourself?

How many Bardo Safaris are you running at a time? There are dozens of those same things running side by side with each other, dozens of similar realities and totally dissimilar realities all running at the same exact time. You don't have to keep track of them when you're in the psychic intuitive, but you do have to balance and harmonize all of them somehow.

In the Afterlife Adventures, in talking with a Blackjack dealer, you are in fact having a conversation with a Blackjack dealer, but at the same time, in the same conversation, you are speaking with a spirit guide, and they will speak to you and you're getting messages both from the dealer and from the spirit guide.

I know you'll remember this, and Jazzy, I know you were there for this thing, and a number of other folks were there at the same event when David, who was one of the casino bosses, was Blackjack dealing to us, and suddenly he started talking what we would call "work ideas." It was so clear and so obvious and everybody went "Ahhh," because he was straight out with it.

St. Mike and I were walking down Hollywood Blvd from Vine Street and Mike says, "I wonder which Bardo we're in right now?" We got two or three more blocks toward Highland and a family is walking toward us; the little kid is maybe three years old and he looks up at Mike and he says, "You're in the third Bardo, Mister." And they walk on. Just flat out. "Thank you," he said, totally unsurprised, because he was a very experienced voyager. It's there all the time.

E.J. Gold experimenting with crystal radios to make highly specific controlled-frequency Beta-Blockers.

Claude and E.J. surrounded by hundreds of experimental components and dozens of crystal radios.

Chapter Twenty-nine

MEGA-WAVE BROADCASTING

Yes, it's happened already, only a few weeks into the whole Brane-Power project . . . I've figured out a way not only actually to tune into parallel universes, but to broadcast open-band into them.

The key to broadband broadcasts directed toward parallel universes is a simple relatively crude early radio technology. It can't be done in FM broadcast, which is line-of-sight Frequency Modulation — it has to be AM, Amplitude Modulation, with the usual skip, bounce and standing-wave propagation syndromes.

Here's how it works: I don't know.

Here's what it does: I haven't found out yet.

Here's what it means to you and me: I don't have a clue.

But even though I don't know much about it yet, and it's only a matter of time before I get this device into the hands of some fellow physicist experimenters who undoubtedly will come up with answers and improvements beyond anything I can come up with, I do have a few simple answers.

I think that what happens is that my carrier signal "rides" on quanta, electrons popping in and out of a number of universes, probably the 11 or 12 basic building block universes, more or less in concordance with String Theory, only I have my doubts about the math involved — I leave that to the Brainiacs who can do the math; I'm just a Theoretical Physicist with a penchant for Thought Experiments and a talent for Circuitry of the Third Kind, meaning vintage crystal and AM technology.

So . . . what I think happens is that my carrier wave rides on electromagnetic "quanta" into a number of neighboring universes within a few billion "Kakus" of frequency "distance" from our HomeWorld Universe.

It then returns via the same carrier quanta, and we get the modified signal back to us with some changes, some minor or major alterations, from the Parallel World transmitters that I've built or someone else has built in those other dimensions.

The result is very weird broadcasts from a variety of other worlds. Some of those broadcasts are talk shows speculating on what might have happened had the South NOT won the Civil War, or had Rome NOT survived into the 21st century!!!

The music alone is worth the price of admission. Jimmi Accardi and I have tuned into a number of alternate universes to make our songs, and it really saves a fortune in song rights, because they haven't yet worked out a way to share royalties across Parallel Worlds, although with the advent of devices like the ones I'm working on right now, the lawyers will undoubtedly find some way to enforce cross-dimensional BMI and ASCAP regulations.

What YOU can do:

You can set up one of my MegaWave Broadcast Units in your Brane-Power Training Center and broadcast throughout your training center the resulting cross-dimensional programming.

You might want a more powerful unit, which is my MegaWave Broadcaster Mark IV, which broadcasts at a radius of 2 miles (that's four miles across!) and several billion Kakus distance across the Parallel World dimensions!

MegaWave Broadcasters are on view at yoyodyneindustries.com

and are a product of Yoyodyne Propulsion Systems, a growing excited company from the 37th Level, my originating HomeWorld, where I actually live, if living is the word for what I do there.

At last, you can broadcast LIVE across billions of Parallel World dimensions, reaching an audience of God-Knows-What types of lifeforms in a variety of Who-Knows-What worlds with God-Alone-Knows physics and cultures!

Be the first on your block to broadcast LIVE into other dimensions! Contact me at the 37th Level NOW, on my contact form at gorebagg.com and I'll give you all the details on how you can MegaWave Broadcast to the weirdest alien lifeforms that ever populated a growing group of Reality-Membranes and Cosmic Universal Strings and you can start doing it right now, TODAY!

Carlos broadcasting into alternate dimension L315a live from Planet Earth!

Ultimate coolness is a Beta-Block amulet in a Parallel World!

Chapter Thirty

SWITCHING TRACKS

If you had a way of avoiding plague or a giant asteroidal impact, would you take it?

On some Earth Tracks, the planet endures terrible disasters. On others, a much lighter, less deadly event might occur, and on still others, maybe harder to reach but still attainable, those disasters don't happen at all.

One of the problems with switching tracks is that although those disasters don't occur, there could be elements that don't match your criterion for survival, such as an Earth Track where humans don't exist. You don't get the Swine Flu there, but on the other hand, there are no humans to catch it.

First of all, understanding what happens when you switch tracks will make a profound difference. Finding the right track to switch into is obviously very important, and survival might not be at the top of your list if other things don't happen as well, such as personal freedom.

In one Earth Track, there is no Swine Flu, no massive earthquake, no Yellowstone Caldera burst, no giant asteroidal impact, but on the other hand, the United States is under a military dictatorship after the present government fails and the major industries topple.

In yet another scenario, although none of those disasters occur, on the morning of January 27, 2018, a nearby star goes supernova and the Earth is bombarded with a burst of gamma radiation, wiping out all life down to the very last bacterium.

A very different scenario awaits one in Earth Track L317a, where everything else is fine, except that China, Iran, Pakistan, France and Libya have entered into an agreement which basically wipes out all competition, and the United States and basically the whole Western World is owned by this very determined and aggressive Power Bloc.

Are there any Earth Tracks that offer personal freedom and no disasters? Not really. The absence of personal freedom is often tied to enormous international and natural events. But there are options in which, although the natural disasters are still present, they are very reduced in threat, and personal freedoms are preserved amongst the smoke and ruin.

One of those primary Alternate Worlds is L315a, which offers refuge to those suffering in the Earth Tracks that are under natural or political attack. One of the problems before us is that L315a is very hard to reach and is quite distant in frequency, squeezed between two massive radiation beacons that hide it in the shadow of their enormously powerful radio broadcast waves.

How can we reach this haven? First of all, not everything is terrific at L315a; there are still problems. It's not a free lunch by any means, but it is far better than what is in store for you if you're reading this.

Getting there is half the fun. It takes a bit of maneuvering to get close enough to the target to make it all the way into that dimension, something like making several drives to get on the green in a game of golf, then sinking a two-foot putt.

Another way of understanding the process is to see what happens when docking a space shuttle at the International Space Station. The first problem is getting into orbit at all, just leaving the Earth's gravitational field. The second problem is getting close enough to the orbit of the space station to make docking possible.

The next issue is to actually micro-maneuver into docking position, then micro-managing the approach in order that neither vessel nor station are knocked out of orbit or damaged.

Then there is connecting the airlocks, and finally actually disembarking from the shuttle into the station. Getting back home is yet another even more complex and dangerous series of events.

When considering a safe haven such as L315a, one must consider issues in navigation and close-targeting — what we call "coarse tuning" — and only when the target is in actual near-sight range do we begin the process of "landing" within its field.

What happens when we get there? As you are no doubt well aware from many science channel specials, particles are always knocking about from one dimension to another. What's to prevent us from getting knocked back into an even worse dimension than the one we ran screaming from in the first place?

One method is to continually travel with a locking device that keeps us in the same track. This device is called the "Black Diode Tracker" and it is intended to keep us on the same track, to minimize "frequency-drift". Can drift still occur? Yes, it can, but with other tools to keep yourself on track, that drift can be minimized if not entirely eliminated.

One such tool is the daily use of the Beacon. This ensures that, even if drift has occurred, as it must in a universe subject to quark immersion and electron sub-orbital quantum mechanics, the subject — which is you — can make continual course corrections back to the target safe haven.

Another device which might help to keep you on the right Earth Track is the MegaWave Broadcaster. You can mount one of these puppies in your home and it will broadcast a continuous carrier wave tuned to L315a. You can order one with a broadcast range of up to 2 miles in radius, but you will have to install an antenna mast for this to be effective.

Within the range of a home, a much smaller antenna, even an indoor antenna, can do the job with a good ground and some care in the installation. A Brane-Power center will naturally have a MegaWave Broadcaster at its center, and if you aren't up to the task of installing one, a Brane-Power operations engineer can put one in for you — in fact, I highly recommend that course of action for a professional Brane-Power Center anyway.

Another effective weapon against frequency drift is the Pyrite Crystal ring, and I'm continually working on ways to combat reality-drift events, including bath salts, personal fragrances, framed paintings containing radionic material, tracking Beta-Block bracelets, and even t-shirts with powerful radionics embedded directly into the heat-bonded patch.

Will all of these work? Together they can work, yes. Will they work without your active participation in tracking and keeping on track? Probably not. It will take a daily practice and application at the Beacon every single day to prevent automatic drift from bringing you right back to the disaster area from whence you came. Without your daily work, you can expect to drift right back into hell.

Is all that work worthwhile? Depends. How much do you enjoy hell?

You don't want to drift right back into hell, do you?

Chapter Thirty-one

STABILIZATION WITH BETA-BLOCKERS

Before we can really assemble our para-selves into a coherent cluster, we need to stabilize in one relatively safe spot, and keep ourselves from wandering in the Bardos, drifting aimlessly from one reality level to another.

Wandering in the Bardos is actually a condition in which the Essential Self continually drifts through the Frequency Reality Spectrum from one world to another.

Not surprisingly, the net result of this wandering is that the world keeps changing its nature to one degree or another and the Essential Self is never quite able to assemble its scattered selves into a single manifesting entity, thus never achieving much attention, presence and will.

Without these three Essentials, the Self cannot possibly do very much about its condition and certainly without some "weight" of self-ness, the Essential Self has no center of gravity with which to pull in its scattered selves from a variety of frequency habitations throughout the Reality Spectrum and bring them into one single manifestation.

The situation is this:

1. The Essential Self has no power without The Assembly. It needs The Assembly to give it sufficient "weight" to become a gravity-well for the self.

2. While wandering in the Bardos, meaning the variety of realities on the Reality Spectrum, there is no central place of congregation, no single focal point to which the scattered selves can go.

3. As long as the Essential Self continues to wander, it cannot possibly concentrate its attention, presence and will for any task, let alone the task of self-assembly.

Given this condition, it is necessary to first stabilize one's position in the Reality Spectrum, and only then begin the process of self-assembly. The practical side of this is to project the attention into a single Reality Frequency such as L315a and keep it there at least long enough to accomplish the main purpose, that of Self-Assembly or what is sometimes called Self-Realization or Self-Invocation, The Gathering.

And while we're at it, why not make it a Safe Haven, which is why we try to keep ourselves tucked into L315a, which although it has some pretty difficult events, is not nearly as bad as some realities where radical catastrophes abound, and the potential for Self-Invocation is taken away by natural disasters, political forces and cultural taboos.

Along with all the electronic methods for hitting and remaining in L315a, keep in mind that for many decades I have produced a variety of “magical” or “mystical” tools for the same purpose, expressed as an Altar Collection. This powerful vibrational tool should not be overlooked in the rush toward techno-tools; they both work together and feed reciprocal energies as a whole, in synergies much greater than the sum of the parts, similar to the structure of a Whole Protein.

So here is the plan, made simple:

1. Find Safe Haven, L315a, and work to remain there as much as possible, correcting frequency drift every hour and every day as you are able.

2. Place a variety of “course-correction” tools around yourself as you are able to — you might find references all along my time-path in a number of books, talks and artworks to the concept of course-correction, such as the Mala Handbook and the so-called

"Lost Works", a group of pamphlets and broadsides, which are now contained in a single volume from Gateways Books & Tapes.

3. Wear your course correction amulet and other wearable Beta-Block and Brane-Power gear whenever possible.

4. Place a powerful beam in your home, meaning the Mega-Wave Broadcaster, and keep it operating as much as possible.

5. Use the recorded material that I've provided, to keep yourself focused in a single frequency on the Reality Spectrum. These are all available from onlythebestcds.com and Gateways Books and Tapes — gatewaysbooksandtapes.com

6. Place Stabilizer Artwork such as my Thangkas in at least one room in your home, and if possible, place them in your workspace as metal framed miniature desktop artwork, using the recommended special metal frames.

7. Carry a Black Diode keyring, moneyclip, etc. as you are able.

8. Stay tuned to GorebaggTV, even if you turn down the sound and don't watch it actively, to keep the Beacon on yourself and your space.

9. Keep your emotions to yourself; don't display them, act out or give way to negativity.

10. Give yourself plenty of room, meaning take the time each day to stabilize yourself on whatever Beta-Blocker you are able to obtain. If you can get on a Beacon for two hours a day, that would be ideal, either at a Brane-Power Center or at home. If you are able to set up a Brane-Power Center, it makes the task much easier and it can eventually pay for itself if you allow others to come to your center for Beacon Runs.

11. Get on the training program, attend the weekend online workshops and stay tuned-in to the L315a wavelength in that way as well, which will give you a stabilizing force and at the same time the opportunity to be aware of developments and to ask questions and get guidance as you work toward Self-Completion.

E.J. with Beta-Blocker Weekend Warriors Across the 11th Dimension!

Across the 11th dimension with E.J. Gold!!!

Chapter Thirty-two

THE BEACON AND THE CHRONIC

The principle reason you don't use the induction without the Beacon is because you want to allow the Beacon to disarm the Chronic. That way, there is no obstruction to your Parallel World voyaging.

One possible worthwhile objective in your Beacon appointments is to sit without the Chronic for a while, to get used to functioning without it.

The Beacon has two main functions: one, to bring two or more mis-aligned Parallel Universe Personas into a single unified manifestation of the Essential Self, and two, to disarm obstructives, the most deadly of which is, of course, the Chronic.

Editor's Note:

For a full introduction to the Chronic as an element of transformational psychology, read the author's book, The Human Biological Machine as a Transformational Apparatus.

Using the Beacon Automapper to track the alternate self in its many diversified Parallel World manifestations.

To those with shamanic skills already developed, the Beacon Automapper makes it even easier to cross the interdimensional barriers!

Chapter Thirty-three

HOW THE MAP WORKS

The MAP is a series of case-specific Parallel World "tracks" that you'll want to follow to begin your journey into PUP Clustering.

What this means to you is that you'll be using my induction CDs for the first 30 Parallel World General Field Contacts, which are the basic contacts that everyone will have to connect up before proceeding with Special Field Contacts that are unique to each individual Voyager.

After you've stabilized in your first 30 Parallel Universe (Bardo) Contacts, you'll be sitting with the Beacon Mk I or Mk II on your own, meaning that you'll now be entering a series of highly individualized "Special Field" Parallel Universe Contacts.

Keep in mind that when voyaging in the Bardos, Parallel Worlds tend to be interpreted as "Between Lives" states, when in fact you only have one life-line which has no beginning and no end, just stopping points and a variety of singularity tracks which read to the Voyager as "lifetimes".

It may seem to you that you are traveling between worlds, but what is actually happening when using the Beacon is that you are merely connecting yourself, "hooking yourself up", to a number of parallel lifetimes that are proceeding right now.

Some of them, be warned, may not be human. Some may be in

universes so very different from the one that you've become familiar with that the physics is unrecognizable and strange. Some of your parallel lives on the other hand are almost exactly the same as the one you're identified with at the moment, with perhaps one or two small differences, such as in one world you die of swine flu or extreme poverty, and in another you get mildly sick with swine flu and your income level is within the comfort range.

By the way, one way of handling poverty is to simply change tracks to a prosperity track rather than trying to bang your head against a brick wall of labor reductions, firings and the epidemic of down-sizing and off-shore out-sourcing of businesses in this tough economic environment.

One clear and easily obtainable high-prosperity track is at L315a, which is the very first induction I send you when you receive your Beacon. This should markedly improve your situation, which you will notice will profoundly alter your life situation and environment. When you hook up to your parallel self in L315a, look for an immediate change in your prosperity level — this is a good indication that you have successfully made that hookup and got it stabilized, balanced and fully integrated in your Essence-Manifestation.

The next several inductions will be toward Parallel Worlds that offer slightly less environmental threat. These will require a change in crystal grips from quartz crystals to carbonaceous chondrites.

After that, we'll be heading for Parallel Worlds with improved health pictures, fewer social and economic pressures and less personal friction. This area will be traversed with crystal grips of the amethyst variety.

Following those minor improvements in your life situation, your work efforts should be much stronger, with fewer interruptions of service and less stress on you to keep up with the Kardashians. You might even see a noticeable side-effect — a definite reduction in bad daytime and sunday television programming, but don't expect commercials to get any better . . . they don't.

Over the mid-level inductions, we'll be looking to hook up to those Parallel Worlds that offer better Essential Self work conditions.

Then in the last leg of the inductions, you'll be working toward conducting your own course of voyaging in which you'll be going Out There alone but with high levels of protection. It's at this stage that you'll want to change your crystal grips over to Black Tourmaline for protection against malevolent disembodied entities and, yes, they're out there just as they are in online videogames.

Now at the point where you've completed all the General Field hookups and they've stabilized, harmonized and integrated fully and completely with no occlusions between personae, you'll be ready for your first voyage on your own.

Not to worry, the Beacon AUTOMATICALLY seeks the next-nearest "hot lead" and goes right to it, totally without guidance from you . . . as a matter of fact, the homing-in procedure happens so fast that you haven't time to actually assess the situation — it's all over before you know it. These Lone Voyages will be conducted on your Quartz Crystals once again, just as the first ones were.

You'll be using your other crystal grips as you go — they come into play depending on what you run into Out There.

Your very first induction will show you how easy it is to hook up your Parallel World selves and Self-Invoke yourself into a high state of Being and Essential Self Presence!

Wishing you great success on your very first Parallel World Voyage!

Experienced voyager George goes far out on a Parallel World exploration, Neal St. Center, 2009.

Glam is what shines from within — Om Mani Padme Hum — hail the jewel within the lotus.

Chapter Thirty-four

WHAT SECRET DID THE ANCIENTS POSSESS?

Have you ever wondered why someone who is not particularly "handsome" or "drop-dead gorgeous" is so attractive to others?

Do some folks seem to have automatic "charisma" or charm?

Did you ever consider exactly how it came to pass that some people seem to have incredible powers of persuasion?

If you examine the people who seem to have the most power, the greatest influence and are the most attractive to others, you won't be able to pin it down . . . . They're not especially good-looking, and in some cases, downright strange or funny-looking. You can easily think of several examples of people who rose to great power or to instant stardom, definitely not for their good looks.

Glamour has nothing to do with how you look. Your appearance counts for only one or two points on the Glamour Meter. What really counts is what's inside, what's radiating out from your inner being.

There are some folks who can walk down a street and men, women, dogs, children and horses will try to follow them home. Why is that?

It's simply something called "Glam".

What is Glam?

It's an internal concentration of positive energy combined with a powerful magnetic center, caused by the accumulation of spiritual force.

Try just 3 minutes on the Beta-Blocker and see the instant result. The accumulation of Spirit Radio Force will amaze you.

How does the Beta-Blocker work?

1. Beta is always with you. It is the result of years of conditioning to use the Beta Brainwaves natural to the process of human thought. Beta Brainwaves suppress Alpha and Theta Brainwaves, which are associated with the creative force, the psychic-intuitive powers of the natural mystic.

2. Parallel Universe Selves tend to accumulate where the psychic-intuitive is focused. That means more "You-Power" as the many selves in your Parallel World Map start to collect in your enhanced manifestation, producing a Power-Presence that automatically gives you greater charisma.

3. When your Glam-Power is higher, you feel more confident. When you feel more confident, your self-esteem goes up several notches, and that shows on your relaxed face and posture. Tension makes you look older, feel older and less confident. Beta-Blocking brings back to you that childlike sense of wonder and magic about the world.

It's as simple as 1, 2, 3 — Beta-Blocking makes you more glamorous and confident, and you'll feel better about your decisions, your relationships and your abilities and skills, in just 3 minutes a day!

Chapter Thirty-five

WORMHOLES

Everyone today knows about wormholes — at least they know that wormholes exist and that they can be used to connect widely separated points within a universe and between universes, even distant frequencies of universal "branes".

What most scientists today are unaware of is that wormholes exist virtually everywhere in space and time and that we pass through them several times a day and even several times an hour under certain circumstances.

Wormholes are scattered throughout space and time, and are very much like meteor swarms in that they tend to swirl in disk-formations and that they tend to gravitate around a black hole composed largely of dark matter.

One factoid to keep in mind is that physicists have not yet quite gotten to the point where they say, "Oh . . . so e REALLY DOES equal mc^2!!!"

What that really means is that energy really really IS mass times the speed of light squared. That sounds rather obvious to most physicists, even those of the quantum persuasion, but it's somewhat like getting the full realization that it really really IS all in the Phase Factor, meaning that just because you've reduced the equation doesn't mean the Phase Factor has actually "gone away".

In trying to grasp how Wormholes can be used for interdimensional voyaging into the Between-Lives State — which is nothing more than alternate dimensional travel — it must be understood that most "aliens" visiting the Earth are not interstellar or intergalactic travelers . . . they're actually coming from radically different alternate worlds, dimensions and universes.

How they get here is the same as how you switch tracks between Parallel Worlds all day long, except that they understand the rules of engagement, and most humans don't yet know how it's done.

There are a few exceptions. Mostly they're what are called shamans, witches, warlocks, ghosts, demons, angels, psychic mediums, astral travelers and weirdos, by those many billions of humans who are "land-locked" and "Earthbound", and who are terrified of the prospect of leaving their precious home-soil, even though, unknown to them and unrecognized by them, they do it all the time.

Why don't people know they're switching reality tracks all day long? Because they've been conditioned not to, by the capping of the Beta-Brain upon their natural Alpha-Theta rhythms.

Up until you were about five years old, when you started "school" conditioning into the Beta-Brain world view, you saw and understood that you were space-changing all the time. The sheer magic and wonder of it all did not escape you.

Then it became time to "grow up", to give up the magic and the sense of wonder, so you could become a useful cog in the machine of military-industrial human existence.

You were taught to get up at a time you would have preferred to sleep, to eat something you didn't want to eat for breakfast, to dress in clothes you didn't want to wear, to get into a smelly, gassy bus that reeked of years of careless usage, to travel to school with unruly kids who teased you if you were even slightly different, to study things you were totally uninterested in and to play at games you hated, while waiting for that precious weekend break during which you were forced to do homework when all your friends were free to enjoy the sunshine and fresh air.

Once in a while you were forced to endure a family excursion with a dad who ground his teeth and snarled at the endless line of unmoving honking cars in an ocean of gas fumes and a mom who urged you to be quiet and sit still while you waited to go to the bathroom and get some treats.

What was all that about, anyway?

It was to train you to join the work force, to labor uncomplainingly — or to complain, as if bitching about it would ever do any good. You were being conditioned to work for the state while holding the belief that you were really working for yourself and your family. A moment's examination will reveal that you were never free to be what you wanted to be.

The illusion of freedom keeps you a slave.

You don't tend to struggle when you think you're free. But look carefully at what you have become for the "good of the state", and you'll note that you are really just a tiny cog in a somewhat larger wheel in a much larger machine we call the human race, which has at the finish line the promise of death.

Resurrection is the bait that lures you to keep on going, round and round, on that grinding wheel, with a heavy millstone on your back.

The Powers that Be would like you to never wake up. Oh, one out of a million waking up is okay — that upsets nothing in their precious apple cart. But what if millions or billions woke up? That would never do.

Now you're ready to hear about the conscious use of Wormholes.

Wormholes are always there. We run into them in the same way that the Earth runs into meteor swarms, which is that they remain more or less still, but relatively, they are in our path as we travel around the sun and the sun travels around the black hole at galactic central, in a Roche's Limit controlled spiral rotation.

What this means to you is that you enter Wormholes all the time. To control what happens to you as a result is merely a matter of seeing the crossroads of alternate universes — Parallel Worlds — that open up at any given intersection of realities.

At any point where there is a choice, there are at least two Parallel Worlds open to you, two or more paths to walk. The choices must be seen to be consciously selected, otherwise you'll just blunder around, wandering in the Bardos forever lost and frustrated with the outcome.

In order to see the choice-points, you need to re-train yourself out of the Beta-Brain and back into the psychic-intuitive state of the Alpha-Theta Mind. This can be easily achieved with any Beta-Blocker technology, but the Beta-Block training can be combined with Parallel Universe sensing. This is what it means to "become as a little child".

Sensing is the first step toward recognition that a choice-point between Parallel Worlds is at hand. This is what it means to "die before you die".

Knowing which path to choose is yet another stage of development in the Essential Self's voyage toward total enlightenment. This is the real meaning of the phrase "Know Thyself".

Understanding how to actually walk the walk, how to be in "Every-Minute Zen" and how to travel the Path of Enlightenment is what is meant by the saying, "The Tao that can be followed — meaning the cattle path — is not the Tao".

There's more, much more, to the technique of walking in the Zen of the Tao, the Path of Enlightenment, but it all has to do with maintaining the Thread of Consciousness and a constant work-path as the ever-shifting sands of Parallel World Realities change and swirl and flow.

You cannot consciously walk a Path of Reality unless you can see it and you know how to take the twists and turns, and that's what Brane-Power Training is all about.

The cattle-path is not the Tao.

Chapter Thirty-six

BETA-BLOCKER MEDICINE?

No, the Beta-Blocker is not a medical device. It is not, nor was it ever intended, to diagnose or heal any medical condition. I am not nor was I ever interested in "healing" of any kind. The body holds no interest for me. I am strictly focused on the Spirit and its voyages through space and time in the pursuit of True and Total Enlightenment.

Healing for me is strictly the province of the professional healer; this can be allopathic or homeopathic or Christian Science for all I care. My mission has nothing whatever to do with healing and if you read my thesis on Beta-Blockers and Parallel Dimensions, you'll understand why the subject of healing never comes up.

My devices are intended strictly as "Guideposts" or helpers in the direction of understanding how to navigate one's Path of Total Enlightenment. They are helpful in my opinion in the opening of natural non-chemical ways to deepen the state of meditation, the expansion into Cosmic Consciousness and the understanding of one's place in MANY UNIVERSES, not just one.

The use of radio waves is only incidental in my technique of deepening the meditative state and combining meditation with a daily discipline of work, play and rest.

Walking the Path Between Dimensions is the province of the Spirit. Healing is the province of physicians. The very word "physician" limits by definition the scope of a doctor's interest — the domain of the body. My interests are totally in the realm of the Spirit.

It is true that a healthy and happy Spirit will influence the body to some degree. I am not certain that this is a good thing, it is just a known fact. I am not at all sure that good health is a positive in the journey — perhaps the trials of poor health are good for some Spiritual journeys. I leave this question totally in the hands of the philosophers. My path is that of Interdimensional Traveler on the Path of Total Enlightenment, and my mission is clear: help all those who would travel this path to find their way onto it, and to navigate themselves to Total Enlightenment.

So let's once and for all settle the question: is the Beta-Blocker or any of its derivatives such as the Beacon or the Black Diode a healing device? The answer is emphatically, "NO!"

Healers can cleanse with the Deep Delta but it is not used for the healing itself.

Chapter Thirty-seven

ASTRAL SLIPSTREAM

Most "aliens" are interdimensional travelers. The ones who aren't are wormhole navigators.

In advanced civilizations — which Earth isn't — "sailing" the wormhole seas and oceans is just as common as sailing the seas and oceans of Earth are to the present Earthian cultures.

It wasn't that long ago that Earthian humans were dumbfounded by oceanic travel. Trans-Oceanic shipbuilding is a fairly recent science dating back to the 15th and 16th centuries, although small oceanic vessels were built by the Vikings, Phoenicians, Minoans, Atlanteans and Polynesians as early as the 3rd Millenium B.C.

Still, it wasn't until recently that ocean voyages for the average person became commonplace.

Air travel is of course even more recent, and when I was young, there were no jet planes, only noisy erratic prop planes which flew at relatively low altitudes and were totally subject to weather and bumpy air pockets, and had crude instruments for navigating fog and for landing.

Now, air travel is quite commonplace and distances are cut by the increased speed of air traffic. Instead of hundreds of flights per day, it's in the tens of thousands. Planes are so computerized that they almost fly themselves . . . and we're just on the brink of space travel not just for astronauts but for the common folk — that would be you and me. Some civilians have already flown into space, at least into orbit.

Many ET civilizations have survived their own Nuclear Age. We might not, because humans are very warlike creatures, just like our nearest DNA neighbors, the chimp.

Will there ever be a nuclear war? We already had one. It was called World War II, and during that war, two nuclear weapons were dropped in Japan, one on the city of Hiroshima and the other on Nagasaki . . . and what's more, the Germans had The Bomb and were just on the verge of exploding it in the harbor of New York City when the war ended.

Just as there are many kinds of "aliens" or ET visitors to Earth, there are many modes of interdimensional travel. One of them utilizes a simple survival vehicle which humans have characterized as a UFO or "Flying Saucer".

The shapes of UFOs will vary considerably and many of them are not at all saucer-shaped. Some are cigar-shaped, some are shaped like a spinning top and some look like strange lumpy rocks or meteorites. Still others are unclear and tend to pop in and out of Earth's dimensional atmosphere like electrons vanishing from one orbit and suddenly appearing in another location.

Methods of travel will vary as widely as their vehicles, but when they appear on Earth with UFO, USO or UBO vehicles, you can be sure they have made their Trans-Dimensional Sea Voyage by way of wormholes, most of which are created by the vehicle's "warp engines" as opposed to their antigrav local motors.

Interdimensional travel depends largely upon interdimensional navigation. Some ET visitors know how to get back "home". Some don't, and in this way resemble the Vikings who had no plan to return once they got wherever they were going.

As civilizations on Earth learned more about sea voyaging, the science of navigation also grew and expanded accordingly. Cartography — map-making to you — and the use of astrolabes, compasses, depth sounders, anti-weather measures, long-distance submarine travel and more became urgent. Accurate time-keeping is a product of marine voyaging.

Einstein explored what happens to time during lightspeed travel. His whole exposition on a Unified Field of Reality is dependent upon what he imagined would happen if someone were able to run with photons at the speed of light, and much of interdimensional travel is about answering the magic question, "If I were traveling at the speed of light and I turned on a flashlight, would anything happen?"

There are several types of interdimensional travel that you need to be aware of. One of these is IBV, or In-Body Voyaging, in which one enters a UFO vehicle and navigates through one or more wormholes from one dimension to another or through a single "Brane" dimension by use of wormholes to leave and re-enter that universe.

We should differentiate "universe" and "dimension". A universe is a contained space-time continuum boundaried by a membrane, or "Brane", while a Dimension is merely a vector related to locational loci or "points" of place-ness.

When traveling In-Body from one dimension to another, definite space and time phenomena will be experienced. Time happens differently as we travel up and down the scale from energy to matter, from e to m in the equation $e=mc^2$.

But . . . there is a different style of interdimensional travel, one which does not depend on UFO vehicles and which has been used by shamans and magicians and witches and warlocks and all sorts of religious nuts throughout the ages of Earthian human existence.

This "OOBE" — Out Of Body Experience — type of interdimensional travel is commonly called "Astral" travel and is quite common. As a matter of fact, it is experienced by YOU every single night, but you have been trained by your culture to call it "dreaming".

Every single night you leave your body for some length of time. It might only be seconds or even milli or micro seconds, but leave it, you do.

Of course remembering what happened is not easy, because in order to leave the body, you must pass through the deep darkness of the Delta IV stage of sleep, which has the effect of short-term memory erasure, to buffer your psyche from any off-world or strange experience that might impinge upon your everyday psyche.

The Essential Self can handle many types of existence and many kinds of physics, but the delicate daytime psyche typically can't . . . .

And so we adjust our dreams to resemble our daytime reality.

If you were able to remain conscious during Out of Body Interdimensional Voyaging, which we call “Astral Slipstreaming”, you'd be easily able to remember what you'd seen.

Keep in mind that you travel in this way all day and all night long. You slip in and out of a variety of Parallel Worlds all the time . . . but you are trained to live in the Beta-Brain, so you don't notice the space changes that happen to you all the time.

When you were a child, you had no such mental buffer, and you were aware of the continual shifting realities.

It's time now to re-learn those skills and to see where you actually are.

Until you can see the path, you cannot consciously walk the path.

Shifting realities happen to us all the time.

Chapter Thirty-eight

VOYAGING BY POSSESSION

Being "possessed" has lots of negative connotations among Earthian humans, but it isn't all about malevolent spirits entering our bodies while we sneeze, cough or yawn. We have counterspells against evil spirits coming in through the open mouth . . . "bless you" when we sneeze or cough, and we cover our mouths when we yawn, to block the open doorway.

The fact is that we are possessed anyway, every single day and night of our lives, whether we like it or not. We have a variety of astral parasites, some of which are your younger selves, some are your parents, teachers and acquaintances of your youth, still haunting you with a variety of commands, taunts, demands and beliefs.

We ourselves possess others. When traveling from one Parallel World to another during the night-dreaming state, we occupy a variety of bodies in a number of different universes, and some of those bodies we drift into and out of might not resemble human biped life-forms at all.

In conscious Astral Slipstreaming using a Beacon Strapper, we might be a bit uncomfortable about occupying a lifeform that is very different from the one we identify with so easily and convincingly, but the fact is that with the mental buffer of the Beta-Brain, we cannot really know what our normal body actually might be.

We slip and drift and stumble and wander from one reality stream to another all day and all night long, and as we do, the self-image shifts with the voyage, making each "place" or "body" seem as if it were the one in which we were born or spawned.

Nothing could be further from the truth.

Through the control of the Beta-Brain, all facts about ourselves, our history, our beliefs, our destiny and our present relationships are shifted too. We assume that we have always lived in the body we are in at any given moment, and that its history is our history.

One single moment's examination will prove this assumption totally false.

You are not the person you were as a child, or as you were ten years ago. Memory of that person's life will seem as a dream to you, as well it should. You have not been in that body very long, and within a very few minutes, you will leave that universe or Parallel World and enter into another, where your Beta-Brain in that world will do its best to convince you that whatever organic vehicle you landed in has always been your home, into which you were born and from which you will someday die.

Possession is the means by which we enter and leave bodies. Not all of those bodies, by the way, are biped, human or even organic. Some are made of light. Some are made of magnetics. Some are collections of elements making a "Portuguese Man-of-War" type jellyfish, an assemblage or colony of separate lifeforms making a coherent somewhat cooperative whole, such as a Slime Mold or a Human Being which is, technically, merely a cluster of smaller organics in a cooperative colony, just like a school of fish, a swarm of bees, or a pride of lions.

Traveling from one universe to another by leaving one form and entering another is technically called Voyaging by Possession, and is the main method used by shamans, who often call it "shape-shifting".

We will explore this method of interdimensional travel using the Beacon Strapper and the Beta-Blocker Amulet.

Why should we be interested in interdimensional voyaging? Because we do it all the time, but we do it in our sleep, even when we are awake.

Where are we going?

Why, Home, of course. But when we get there, we want to arrive with all the beingness and conscious awareness we can muster. And that's the point of the Brane-Power Training. To assemble ourselves around the Essential Self and to arrive as a collected beingness, to Invoke our Presence and to truly Be Here Now. Voyaging by Possession is one method of achieving this.

Using Astral Slipstreaming, you can easily navigate to the best work-path and bring yourself into the world in which you are Enlightened and working in The Work. You can only make conscious choices if you are able to see the crossroads clearly. Brane-Power is the way Home.

Walk the Path With Eyes Wide Open.

E.J. and Koyote at the Brane-Power Neal St. Center in Grass Valley, Ca.

The illusion of a single universe is easily broken.

Every moment takes you further and further from your original home universe.

Chapter Thirty-nine

WHICH UNIVERSE IS HOME?

Heraclitus the Greek Philosopher said, "You cannot step in the same river twice." This is emphatically true when stepping out of one universe into another.

You switch "tracks" all the time. You are constantly in flux, jumping and jaunting from one Parallel World to another. The illusion that you are stable and exist within only one universe is a trick of the Beta-Brain, which translates everything with its Babble-Fish process.

The illusion of a single universe existence can easily be broken with simple awareness. It doesn't even take training or untraining to see it. You are not at all the same person you were twenty years ago, ten years ago, last week or even yesterday.

You travel from one Parallel World to another in a series of bodies by the process of occupation or, more correctly, by possession, as a continually voyaging spirit.

The idea of returning to your Home Dimension is as unrealizable as standing in the same water of a rapidly running stream. It just plain can't be done.

Every moment of existence takes you further and further from your original home, whatever and wherever that might have been.

Wanting to find your way back to that original spawning point would be like wanting to return to the womb. To put it bluntly, you've outgrown it.

Beta-Blocker Black Diode Amulet comes in sterling, 12kgf and 14 karat gold.

Chapter Forty

USING YOUR BETA-BLOCKER AMULET

You can use your Beta-Blocker Amulet in several ways right now:

1. Wear it and notice what happens during the day. Write down your observations in your Beta-Blocker LogBook.
2. Answer questions about it from folks who notice it on you. Tell them about Beta-Blocking and Parallel Worlds if opportunity is there and the door is open, as it were.
3. Use the Beta-Blocker Amulet to remind yourself to invoke your presence, experience the Waking State and to work for the benefit of all beings everywhere.
4. Energize the Beta-Blocker Amulet by wearing it as often as possible.
5. Wear the Beta-Blocker Amulet while using the induction CDs listed on Brane-Power under the title: "Beta-Blocker Induction CDs".
6. Wear your Beta-Blocker while training on the Beacon Strapper to reinforce the extra-dimensional contact and to help integrate the Parallel World persona that has been invoked into your Essential Self-Presence. As your manifested Presence becomes stronger, the amulet's influence on your spiritual aura will be greater.

Remember that the Beta-Blocker Amulet is not a substitute for the Beacon Strapper — it is intended as a supplementary spiritual tool toward total enlightenment.

Kristin models her Beta-Blocker Amulet.

Chapter Forty-one

INSTANT PEACE MEDITATION

Everyone today knows the benefits of meditation, but who in these trying and troubled times has the years to devote to mastering the method? Now you can give the gift of Instant Peace Meditation.

The Beta-Blocker Amulet is a magical blend of science and metaphysics. Give the Twin Turbo to your friends and family and experience the joy of Instantaneous Inner Peace and Harmony.

You can start your very own Instant Peace Meditation Center in your hometown with the Beacon Strapper. Three minutes per day on the Beacon could change your life and the lives of everyone who comes in to try it.

Peace, Inner Harmony, Prosperity are what many people seek.

Deep meditation and Cosmic Connection can provide an answer. Tune in to the Schumann Harmonic and feel the breath of true Inner Peace and Freedom. You will feel more confident and more certain of your True Path, and that's what makes life worth living!

Is someone near you tense, uptight and stressed out with little nagging worries? Give the Beta-Blocker and watch the change! You will be amazed at the actual morphing effect of deep levels of Instant Meditation.

Meditation eases the spirit, makes the tensions between spirit and body relax, and with relaxation comes Inner Peace and Harmony.

Do you want to see World Peace in your lifetime? Don't just visualize it, actualize it. Give Beta-Blockers to everyone in your circle of friends. Sell the Beta-Blocker in your hometown. Peace begins at home. Ever-increasing, the Circle of Peace brings us closer and closer to World Harmony and Peace. Already thousands are wearing Beta-Blockers and they are influencing all those with whom they come in contact.

Peace and Harmony breed Peace and Harmony. The ever-widening Circle of Peace and Harmony are yours to bring about. Tell everyone you know and everyone you meet about Beta-Blockers and experience the deepening well of Peace and Harmony all around you!

George wears a special custom-made “Circle of Peace” amulet.

Chapter Forty-two

CHRONIC & BETA-BLOCK

The Chronic will fight like hell when the Beta-Block is used. A sure sign of the Chronic's reaction to Beta-Block is bitching, griping, groaning and griefing, combined with all sorts of defense mechanisms such as "it's all wrong, everything is too loud, too soft, too long, too short, too hot, too cold, suspicious, dangerous, weird, malevolent entities" and the most obvious and conspicuous sign that the Chronic has been activated, "it isn't working, nothing happened", when it's very clear that something did, or the Chronic would never have triggered off in the first place.

If your Chronic is protecting you against the Beta-Block, don't worry — it will fall off by itself after a while. Meantime, endure it, suffer with it. Don't forget, it's been with you a long time, and like any partner, it won't easily let go.

Your logbook is a powerful ally in the struggle against the Beta-Mind.

Keeping good records of your Parallel World explorations is very important.

Chapter Forty-three

ON THE IMPORTANCE OF JOURNALING

As we are seeing, the chronic seems to be kicking in after three or four significant Beta-Blocker episodes. As they say, YMMV (your mileage may vary). However, all in all, the general pattern is three or four significant Beta-Blocker experiences and blamm — here comes the Chronic rearing its ugly (demeanor challenged) head. Each of us have our own unique brand of chronic. But one thing remains the same — its purpose. Whether you adhere to the angry type of Chronic or the weepy type of Chronic or the overly objectifying type of Chronic, the aim is always the same — to interfere with whatever has proven itself effective at awakening the machine.

With the advent of the Beta-Blocker technology we now can put a name to that aspect of the machine that it is so interested in protecting — the Beta-Brain. The precious (my pretty precious) that the machine is protecting is none other than the Beta-Brain.

The Beta-Brain is like a governor on a golf cart. In the case of a golf cart, the governor is used to limit the flow of gasoline into the carburetor so that the cart does not exceed 25 mph. In the case of the human biological machine, not only does the Beta-Brain keep you below the speed of light it will also limit the otherwise uninhibited functions of creativity, imagination, curiosity, adventure, and other equally intrusive things. Intrusive to whom or what? Well, intrusive to the military industrial complex for one. Art, imagination, curiosity, and even truth-telling are very disruptive to the smooth function of somnambulist droids required by the 9-5 regimented work force. There is a reason that aboriginals have a stronger tradition of shamanism than let's say your standard w.a.s.p. (white anglo-saxon protestant). They are not as plagued by the Beta-Brain as your average civilized individual. Aboriginals don't manage to escape the Beta-Brain completely. But, let's face it, some cultures are definitely more involved in the process of enforcing Beta-Brain adoption at young ages. This typically takes root by the fifth year. By the time a child is five years old they are well on their way to a well-developed Beta-Brain.

So why is it that the Chronic jumps up after three or four successful Beta-Blocker experiences? Well, why should it bother to interfere with your desire to sit around holding a couple of wire-covered crystals hooked up to a mysterious electronic gizmo? You certainly get involved in enough weirdo activities that this doesn't attract any undue attention from the awakening watch dog. But after three or four successful episodes on a Beta-Blocker or Beacon the cat's out of the bag. It is now painfully obvious that this BB behavior cannot be allowed to proceed unchecked — it is too damn effective.

That is when the Chronic is activated — when something becomes too gosh darn effective.

So what does this have to do with the importance of journaling?

One of the common manifestations of this Chronic behavior is cynicism or sarcasm about the Beta-Blocker process itself. When this happens and your brain is screaming something to the effect: "What's this shit all about?" it would be most excellent to look back upon your own notes and observations. Who better to trust on the subject than yourself? Hence the need for careful, thoughtful, honest, to the point journaling by you. Then when the Chronic jumps up and down wondering what the heck is going on, you can refer back to your notes for ****your**** perspective on the topic — personal experience personally experienced.

This can yield some interesting results.

1) If you ever needed proof that you are not always of the same mind, you will have direct evidence.

2) You can read what you had to say about your personal experi-

ences prior to the recent fit. Thus, giving you a bridge between the two spaces/worlds/"I"s.

3) Give you a little hint that perhaps your current state/mind/emotions is not the sum total of your impressions and opinions.

4) Give you a little clue as to the (non)lasting nature of said states, opinions, emotions. And much more.

The main point is that through the process of journaling you will be able to speak to yourself across the chasm of different "I"s.

Thus, do yourself the benefit of being honest, open, forthright, clear and succinct when you journal. You don't want to wonder about your honesty when reading your material as a different I. You should be able to say to yourself: "No matter what I read I knew that at the time I wrote it I was being true to myself."

And you definitely want to be succinct. The middle of a fit is not the time when you'll be tolerant of even your own rambling journal accounts. It is a good habit to adhere to the following simple rules:

Never hesitate to be direct.

Always tell the truth.

Neal St. Center, July, 2009.

Fear & loathing in other dimensions can bring a voyager down . . .

The Chronic can suggest a wide variety of self-punishments . . .

Chapter Forty-four

FEAR & LOATHING IN ALTERNATE WORLDS

With the advent of the Beta-Blocker, it's time to see quite powerfully the net effect of the Beta-Mind and its ugly, vicious spawn, the Chronic.

The Chronic doesn't actually exist. It's merely a name for how the organic self reacts when confronted with the threat of reality in the raw.

The Chronic has several characteristics that define it and that make it very clear that it has been activated, triggered, stimulated.

1. Everyone and everything else is at fault. The Chronic feels much more righteous if everyone else and everything around it is to blame.

2. Nothing is quite right — the Chronic delights in finding problems and wrongnesses and playing a constant game of "What's Wrong With This Picture???"

3. Nothing works. It's all Epic Fail. Hopelessness and despair are the playthings of the Chronic.

4. There's nothing to fear but fear itself. It's all one big threat out there, and the world is full of malevolent entities putting the subject under constant psychic attack while billions of others go unpunished and unpersecuted.

5. I'm a piece of crap. Self-esteem is at an all-time low, and I should be punished. Self-Loathing is the Way Out. If I can hate myself, maybe people will leave me alone.

6. I'm sick of this, I'm out of here. Quitting is the Method of Choice.

7. Don't pity me, I'm busy pitying myself. There's no one quite as miserable as I am, so don't tell me your troubles, I've got plenty of my own.

8. If I didn't have bad luck, I'd have no luck at all. Nothing is working and everybody hates me; I'm gonna eat some worms.

9. Life sucks, then you die. Nobody ever gives me a break, and nature seems dead set to make me suffer.

10. I hate my life but I want to keep things as they are. If I keep doing the same things I've always done, there won't be any surprises up ahead of me.

If you hate your life, it might just hate you back!

Chapter Forty-five

WORLD PEACE & HARMONY

Don't just Visualize World Peace. Make it so.

Have you ever seen someone walk through a roomful of chaotic people and as they pass, people seem to calm down — for no apparent reason, the room gets quiet, almost as if it had suddenly turned into a library or a cathedral?

You can be that person!

With Beta-Block, Peace and Harmony swirl within you and around you.

You become a spiral of good spirit and powerful love.

The pacification of a roomful of chaos is a direct result of Beta-Block. When you're in the Alpha-Theta State, nothing seems important. You're in the moment and everything is wonderful.

When you wear your Beta-Block amulet, you're not the only one who enters into a state of bliss and harmony. Everyone around you is affected by both the Beta-Block amulet and your inner state of peace.

Integration of Parallel Universe Personas can be verified with "the checker".

Tito says — Integration is simple and always an easy procedure.

Chapter Forty-six

INTEGRATING YOUR PUPS

When your Parallel Universe Personae come into your sphere of manifestation within your organic body, there may be a period of adjustment.

Usually the new PUP integrates within a matter of minutes, but occasionally the Chronic may be activated, because the Chronic resists ALL change.

Wearing the Beta-Blocker Amulet can help overcome the Chronic's natural and automatic resistance to change — the Beta-Blocker is designed to defeat the Chronic.

Once in a great while, the Chronic will seem to win, but within 24 hours, the Integration Process of the new Parallel Universe Persona will take place and your organic self will be fully "Meisnerized", meaning balanced and in harmony with itself and its Cluster of PUPs.

The solution to Chronic Activation is merely to wear the Beta-Blocker and trust the process.

Koyote says: Wear a Circle of Peace — give one to a friend.

Chapter Forty-seven

ACTUALIZE WORLD PEACE

Don't merely wear a symbol of World Peace — Actualize It.

The Circle of Peace Beta-Blocker is more than just a symbol of peace like the "peace sign" which originated with Sir Bertrand Russell's CNVA, Committee for Non-Violent Action in the 1950s, which used an abstracted drawing of a B-52 bomber as its icon.

Your Circle of Peace Beta-Blocker is an ACTIVE AGENT for Peace. It works through radio wave vibrations to bring about a real change in yourself and at the same time exerts a small "sphere of influence" around you, an aura of Peace and Harmony which will profoundly affect your environment.

At the same time, it works to bring YOU into the Parallel World where World Peace actually exists.

It seemed like certain death . . .

Escaping a horrible death . . . but for how long?

Chapter Forty-eight

I WAKE UP SCREAMING

It's two o'clock in the morning. I hear a muffled groan, a cry, a whimper and a threatening, growling sound . . . and then I wake up with a scream in my throat, sweat pouring off my forehead and palms.

I woke up just as I was about to die.

But no, it was just a dream. "Just a dream, a stupid nightmare," I reassure myself, and after a few minutes spent doing nothing in particular, I try once again to drift off to sleep, apprehensive that I might end up back there again.

Sleep doesn't come at once. I read, watch a little television, put a DVD in the player and watch a movie . . . then, mercifully, I drift off again into deep, deep sleep, in plenty of time to get enough restful refreshing energizing to face another day at work.

But here it is, four o'clock in the morning, only two hours later, and again I wake up with a gurgled yelp, having narrowly escaped death once more . . . or did I???

It seems to me that death was inescapable, and yet . . . here I am to tell the tale. How can that be???

It's easily understood in the light of Parallel Worlds. At every choice-point and crossroad of life there are at least two paths, each leading to a very different Parallel World, and you just visited one or more of them while your present body slept in Deep Delta.

It's a common occurrence, one which can powerfully influence what we do in the daytime, yet we never really examine our sleep voyages, because we simply cannot remember them.

Now, at last, you have in your hands a potent tool for Lucid Dreaming, the Beta-Blocker Deep Delta. Merely place it by your bedside within a radius of 4 meters, which is almost 20 feet.

No need to turn it on — it's crystal technology, so it's always on, powered by the incredible electromagnetic energy of millions of transmitted radio waves coming from all over the Earth and even from outer space, from near and distant stars.

In the morning, you'll be entering your dream record into your Beta-Blocker LogBook, for later examination, interpretation and correlation with parallel universe experiences.

After some training, you'll be able to remember, and possibly even direct your actions in your nightly dreams. Instead of nightmares, you'll be voyaging consciously in Parallel Worlds, connected by a thread to your present Earthly body, protected by your Deep Delta DreamWalker.

Why wake up screaming, when you can so easily, with Beta-Blocker technology, wake up smiling and happy?

Deep Delta Dreamwalker helps develop lucid dreaming and memory of dreams, plus volition within the dream.

Chapter Forty-nine

MORE TO E.J. THAN MEETS THE EYE

May 16, 2009

How did I actually come up with my Beta-Blockers? I'll start out by saying that in 1964 I was walking down Santa Monica Boulevard, and I was going to be working out of the fencing studio. Right next to the fencing studio was a new-fangled place called The Computer Training Center where they trained people how to use a computer. What it was at the time was IBM 301 card sorters — that's what a 1964 computer was all about. That computer, which took up one block by one block by three stories high, would not even do the work your laptop will do today. Hard to imagine. At that same time my cousin Rick invented a desktop computer which was actually a read relay system with a sorting device. It's just on/off to indicate binary yes-ness and no-ness. Or, in other words, ones and zeros.

From 1962 through 1965 I was involved in super-secret communications gear, which is no longer secret, but I really can't tell you anything about it other than when I dealt with that communications gear, I had to learn certain things about waves and frequencies. One of the things I learned about was called "nulling". Another thing I learned about was wave cancellation. Yet another thing I learned about was atmospheric skip and many other aspects of secret communication. One of the things I found very interesting is that there are some radio waves that travel through the earth rather than traveling through the air.

Human biological machines require some sort of experiential evidence in order to understand something. They can't really understand something that's inchoate, unformed, and so one of the problems with understanding quantum theory is that we have no experiential data to lean on in order to wrap our minds around what we're trying to understand, so as a result we have only hearsay.

The evidence we have for Parallel Worlds is not from the realm of quantum physics, it just requires quantum theory in order to explain it. It's still Einsteinian and Newtonian, little boxes and clickers and counters and photographs. It's still very physical, ordinary Newtonian stuff. We're still looking at traces of things rather than the things themselves.

The Hadron Collider is not looking for direct evidence of extra universal matter. It can't. You cannot possibly see anything quantum, not directly. You can only see tracks in a film or a field of hot gas. We're still at that level where we're looking at Wilson Cloud Chambers and saying, "Well, this moved and that moved; something must have moved those things. I'll bet you it was a *blank*, and it moved it this particular way and it put that particular spin on it."

It's like trying to derive the existence of an invisible cue ball while looking at the break of a pool game. All you can see is the balls breaking, and from that you can derive what the cue ball must have done, what its mass must be, what its velocity must have been, what its acceleration or lack of acceleration might have been, what the collision factors were, loss and dynamics of energy must have been during collision, and so on.

You don't see the billiard balls breaking, you see the track. When the balls hit the bumpers is when you can finally detect them, because the bumper is what your detector gear is all about. You don't see all of the stuff that happens before they get to the bumpers. So you theorize — "It ended up *there*, so it must have started *here*. It must have been collided with by *this thing* going at *this* particular velocity and *this* particular angle of momentum, and it must have had *that* mass, so therefore . . ." but the *therefores* are very tenuous and they're all of those things in the realm of, "We can't really see it." We don't have any direct experiential data to understand quantum mechanics.

Quantum theory is out of reach, it's elusive at this time except, of course, to a psychic intuitive who can tell you, "Well, listen guys, I've been telling you that all along. I told you there were 11 dimensions beneath and there are millions and millions and millions of dimensions above, and they're kind of gelatinous things, like jellyfish, all membrane and fluid and self-luminous and shape-shifting."

If you think of an infinite number of jellyfish, all the same size and all of them kind of flopping around in the same exact space, but separated by frequency, which means that they're not touching each other, but they're in the same space . . . *that's* what infinite universes would look like if you could see them from outside. Each of those universes has its own physics, its own reality, each is convinced that there is no other universe. The inhabitants of those universes may start intuiting that there are other dimensions. But psychic intuitives have seen other dimensions; we *know* that there are other dimensions!

People who experimented with psychoactive substances in the 1960s learned that other dimensions were accessible, but then of course, that was all made illegal because you can't look at those things; you're not allowed to.

As soon as it was discovered that there were alternate universes out there, naturally there was a complete clampdown on the use of anything that might get you there. I have no doubt whatever that at some point somebody in high government will decree, "You're not allowed to look there."

As long as they think that it's frivolous or idiotic or ineffectual, it'll be tolerated. And the moment that something happens to prove to them in their little mindset that, "Oops, these guys have something real," then they're going to clampdown.

As long as it's fringe science we're probably safe. As soon as it becomes real science, and is shown to have military applications, we civilian parallel universe explorers are no longer safe.

In 1962 it was discovered that the Russians were using psychic operatives, *doppelgangers*, to gain entry to very, very secure, very secret, very top secret U.S. Installations. The Soviets have always been very good at stealing secrets, very good at it, and they didn't stop at the physical methods; they went right into the psychic methods because there's nothing in Russia to stop them from looking in those areas because it's *all* classified, it's *all* secret, always has been, always will be.

So, a number of us were recruited to countermeasure those intrusions by KGB doppelgangers. I naturally don't talk much about that. I simply mention that I was involved in the program — exactly what the program was and what kind of results we had and how extensive it was; I will not ever say, but there are places you can look. I'm not going to tell you very much about it. If you look under remote reading you'll find some things. If you go to oldspooksandspies.org, you can find other things, things that I can't tell you.

However, I want to finish answering the question: "How did I stumble across this?" In April of 2009, I was trying to arrange a demonstration of what crystal radio technology was all about from 1900 to 1912. I was very dissatisfied with the plastic piece of crap that we had gotten a number of years earlier that was supposed to be a crystal radio, sold by Radio Shack, and it was total crap.

So I went online and I fished around and found a number of people who had kept crystal radio technology alive for many decades. I found some good people who were serious and who weren't out to make money from this great and powerful long-lost technology. I purchased a number of parts and bits and pieces of things, and so forth. I started meddling around and I came up with a number of interesting crystal radio rigs. The Litz configuration is a very elegant crystal radio, but they're very hard to build.

Then I came across my friend Lance who works at NASA, who said he had a kit that he builds for hobbyists and so I said, "What is it?" He said, "It's a Dunwoody." "Well, okay, let me take a look." He also sent me a Bucher QRM suppressor, and a Morgan antenna tuner which you absolutely have to have when you're working with the Dunwoody.

So, I built the Dunwoody. It wasn't satisfying me. The diode was the standard 1N34A, so he helped me hunt up a Filmore which is a detector trap or a detector cage. Now this was exciting! I had a lump of germanium, then I asked if there was any way we could get pyrite and graphite crystals.

I started experimenting with these things and I hooked up the Dunwoody. It was rather nice. And then I had to put the Morgan antenna tuner in to get a stronger signal, but that loused up my selectivity. So I had to put a string of Bucher QRM suppressors in there to get past all the local stations that were banging out this very heavy overload that spread across the entire AM dial. I finally had those tuned out, and then I could hear at a distance. Now I'm getting signals from all over the globe — and beyond. As I'm sitting there playing with this I noticed that there's a very profound *Beta suppression effect*. I don't need it because I've trained myself a long time ago, a millenium ago, to not use or trust Beta. I don't use it at all, never did use it.

I never cracked a textbook in my entire life when I was in school, didn't need to. I've never studied a musical instrument, didn't need to. I never studied anything that I do because I didn't need to because I can pick the stuff up without having to study it, without having to process it through Beta. I consequently can play guitar, saxophone, clarinet, flute, piano, all sorts of drums including a drum kit down to tambourine, bass and a bunch of instruments that I managed to play without either studying the instrument or studying music or practicing. The same thing is true in all the other areas of endeavor that I engage in, everything from art to physics. I don't need to study it; I just grab it from my parallel universe corresponding self that has the data and experience I need right now.

But I noticed that people were getting a Beta suppression effect all around me as I was working, and I noticed that the effect seemed to taper off at about four meters, after some observation. And I determined at some point that it was that something about the crystal technology was causing this — I wasn't sure just what. So I started to modify — you'd say today hack — I started to hack the Dunwoody more and more and more and more until finally I had a totally different circuit, a circuit that was not even remotely close to the original circuit.

Now I had to have Claude come over and help me trace down a circuit problem. A circuit diagram is different from a schematic. A diagram kind of draws pictures of the things, a schematic is all written in electrical engineering code. And so he helped me to trace down circuitry and to develop circuitry based on electronic theory, not upon guesswork, and he came up with the numbers I needed to get to the Schumann Harmonic, my target for Alpha-Theta enhancement.

So Claude was hammering out these numbers for me, and then he said, "Why don't you just switch these two components over?"

It was then that I realized that this could be a powerhouse of a Beta-Blocker if I only get some more energy out of it, which required that I find some way to get Henrys — which is antenna power from somewhere. I also needed to find a way to improve the circuit so that it had much greater sensitivity and much greater selectivity.

In the 1960s I was one of the first pioneers in biofeedback. I worked with Dr. Hershel Toomim, and of course, my longtime friend, Dr. John C. Lilly.

I came to realize that when we were messing around with biofeedback, one of the things we discovered was wave cancellation. What happens if you take one wave and you combine it with another wave? Whatever the difference is, is what's left over. Now you know that from basic arithmetic. So 20 minus 12 would be 8. It doesn't matter what kind of wave it is. Audio is part of the electromagnetic spectrum, and so is a radio wave, and so is an x-ray, and so are microwaves, and so is light — it's all part of one spectrum. Audio frequencies are just hearable frequencies. There are frequencies up and down from that that you can't hear, but bats can, dogs can, elephants can, whales can. Just because you can't hear it or see it with human senses doesn't mean it doesn't exist.

Another important factoid is that the same exact "note" occurs in very specific intervals up and down the electromagnetic spectrum. So you actually have octaves up and down the spectrum. Gurdjieff didn't have the modern scientifically approved technology to understand what he was saying, but he was right that *these "notes" occur in repetitions* up and down the spectrum. You can make a noise in the audio spectrum and you'll have a corresponding "note" or "chord" in the x-ray area of the electromagnetic spectrum. You'll have a corresponding "note" in light.

MrfixitRick on YouTube.com did a wonderful experiment where

he shows a relationship between light and sound. And by the way, the detectors that I use are so sensitive that they actually can detect 28 shades of gray.

Back in the 1960s, nobody was using crystal technology. Crystal technology had died out around 1912, and by 1922 it was completely gone, giving way to "valves", which you call "tubes". So now we're into the Armstrong tube radio. From tubes we went to transistors, then from transistors to integrated circuitry, and from integrated circuitry, of course, to photoelectric circuitry called microchips and nanochips.

In 1902 to 1912 — the height of the crystal technology — nobody was even considering brainwaves as anything important, or even existing. Nothing was known about Alpha and Theta waves and their connection with psychic-intuitive natures. Nothing was known about right lobe, left lobe. In the 1960s we developed biofeedback. Nowadays you can't go into a hospital or medical center without seeing biofeedback devices everywhere.

But what is no longer common is the crystal technology. Who is building crystal radios when there is digital Sirius radio coming down from the satellites and you've got five trillion stations coming to you all over the planet. So why bother to look at crystal technology? It's ancient, over a hundred years old.

Well, I happen to be that odd combination of crystal technology, waveform technology, quantam mechanics and biofeedback, so when I had something in my hands, I recognized what it was. If I hadn't had the biofeedback data I would never have known what was happening. I would not have attributed the Beta-Block effect to the crystal technology in the room had I not had those strange combinations of technologies that are a hundred years apart — and one technology no longer exists — the crystal technology. You can barely get parts anymore to make a crystal radio.

The other technology is biofeedback, which has now become largely medical. Who in the medical field is interested in using biofeedback in order to explore fringe science such as parallel universes? Scientists at the Hadron will soon know whether they can prove the possible existence of dimensions besides the three that are known today.

Basically they're running an experiment by smashing one thing into another at very high speeds and they're going to look at the results. They've predicted a particular result if they see it, the only thing that would explain it would be that one of the particles involved would have flipped out of this universe into another, briefly — and when we say briefly we're talking not even measurable microseconds, below pico seconds. It's in a realm of time where there's hardly time. And then it flips back. If they see this effect, they will have what for them constitutes proof that there can be a parallel universe. But they could have asked Bea Stanley in 1964. She would have told them there was. She's been there and back.

We have a question from Piph. "Given that radio signals and other types of waves are transparent to one another, would you please brief us on wave cancellation and nulling in the light of that operation."

Who says they're transparent to one another? That's not true. There is photon photon interaction, but for the most part they're invisible to one another — not compositely, they're not.

The gap in this is that as soon as these photons touch matter, they're interacting with each other. So, in the case of the coil in the crystal radio, once they get into that coil they are definitely interacting — they're particles at that point.

The contributing fields are working not truly separately, but on the same medium.

Atoms are invisible to each other. They don't touch each other — they don't actually come in contact with each other. So how can they possibly create matter? This touchable, tangible thing is made up of atoms which don't touch. I can clap my hands and make a clapping sound, but nothing is actually touching anything. It's field against field. The parts don't touch, but the composite waves constantly interact with each other. It's called *oscillation*.

In the kitchen you have five cooks who are invisible to each other. They can't touch each other, see each other or hear each other, but they're all working on the same soup pot. They will interact through that soup pot, and since what they're seeing and tasting is the soup, the composite result very definitely has effect for all of them.

You ask anybody who's in the radio broadcasting industry whether waves interact with one another or not and they'll tell you, boy do they interact! Because they do, there's a thing called the FCC. I've got a 50,000 watt banger over here, somebody else has a 50,000 watt banger over there, we're both on the same frequency. Now my wave pattern better not interfere with theirs and theirs better not interfere with mine. The FCC makes damn sure that those things do not interfere with each other. If you get those radio station towers too close to each other, you're going to get all kinds of oscillation happening in the fringe areas where they beat against one another.

The composite subtractive frequency — the "soup" — is not going to equal the first wave, nor the second wave. It'll be a third entirely different waveform.

In terms of photon-photon interactions, most of what you see comes from coupling through matter of one sort or another. It's called a medium. There's matter all over the place, so these photons have plenty of opportunity to interact with each other in many, many ways. However, that's not even really the point, because as soon as you get into something like radio or any detector, the detector, by its very nature, is made sensitive to those waves, which means they're not at all mutually invisible. Paper is not a great detector of radio waves. Of course it does detect radio waves, but is not a great detector of them.

Now if you take some copper coils and some resistors and capacitors and a diode, that becomes, relative to a piece of paper, a great detector. As a great detector, that means it's detecting not just one radio wave, but all radio waves, and more. Beyond radio waves, depending upon how it's built, it can have quite a wide range of detection. That's the selectivity issue. Just a little piece of copper will be affected by pretty much anything that puts out an electric field force.

Take a Gillette Blue Blade double edged razor blade and a graphite pencil and a safety pin and a little bit of magnet wire to wrap a coil . . . add a pair of headphones, and you've got a World War II "foxhole" radio. The GIs were not permitted to have radios in their foxholes because receivers act as transmitters as well.

It's all about voltage. Voltage variation creates the illusion of sound. It has to be translated into diaphragmatic motion in order for the ear to detect it, which requires that we convert the electrical variations into a mechanical device that moves airways diaphragmatically.

Steve Sachs was a friend of mine who invented the light show. One of his inventions was to take a bass speaker and turn it on its back so the cone was up, and he'd pour mercury into the cone, and then aim a laser down into the mercury, then he'd hook up the speaker to an amplifier. Naturally that amplifier is going to cause that speaker to vibrate. As the speaker vibrates, the mercury vibrates. As the mercury vibrates, it breaks up the angles and changes the angular momentum of the laser beam causing a flickering, darting affect. So the laser beams were flying all over the place, a single laser beam was being broken up in angular momentum. Then he threw that against one of these ballroom mirrors — these mirrors composed of hundreds of little tiny fragments of mirror — and as it turned you'd get all sorts of shooting effects of the laser beam.

What I'm actually exploiting is a micro transmission effect, and the field effect changes that occur from radio waves as opposed to audio waves. Audio waves I have known since the 1960s. When you put one audio wave in one ear and put another audio wave in the other ear, in theory you're going to get a brainwave differential. Actually you don't, and the reason is because you're just *hearing* these things. It can influence you, yes, but it's not actually translating into *whole body oscillation modification.*

Audio harmonics don't have really any effect on the whole body and definitely have no direct effect on brainwaves other than by hypnotics. I would love it if binaural worked as a brainwave device because it's a whole lot simpler for me — we merely produce some audio CDs and send them out — but it doesn't bloody work, because you just have your eardrums vibrating at different rates on different sides of your head. Not your body, not your brain, not your nervous system — just your ears. I wish it did work as a brainwave alteration device because we've got a beautiful audio studio in there and I could just have Oz, our mix man, whip out some binaural recordings. I could have done that 40 years ago.

My job would be a whole lot easier if all it took was some sound waves. They do produce a hypnogogic soporific state.

We hear things through vibration. We don't hear things through electrical variations until it gets into the body, once it passes the ear circuitry, the mechanical force.

The ear mechanism responds to mechanical force, is mechanical force. The transmission of sound through air is mechanical, not electrical, not electronic. Sound travels mechanically through a medium which means you cannot have sound in space. I don't care what you've seen on *Star Wars* and *Star Trek*, you can't have sound in space.

Sound requires a medium to transmit it, but broadcast radio waves do not. You can transmit those through space. The air is vibrating, and it hits my eardrums. The eardrums vibrate with it. The eardrums move forward and back.

It's actually converted to chemical energy. This creates an electrical variation in the nervous system which then travels along nerves and goes through nodes and is retransmitted, rebroadcast through them, and so forth, and eventually reaches the brain. The brain creates a construct of what it thinks the sound was. You don't actually hear anything. You're hearing the brain's reconstruction and synthesis of what it finds understandable. It creates a coherent pattern which you call sound. That pattern is understandable by you, is translatable by you and it's coherent to you. It makes sense to you, but in fact, it doesn't exist as a coherent isolated pattern. It's an internal hologram construct not within the brain itself, but *as if* within the brain.

Here's a microphone. The microphone vibrates with your voice. It creates movement of something. That's still mechanical. That has to be translated to electrical in order to travel through the wire. It travels through the wire as electrical variation. It then goes through an amplifier, storage system, amplifier, comes out the other end and it vibrates a speaker which pushes air. It's like the relationship between a camera and an enlarger. A camera looks through a lens, light comes in in various ways, through a lens and inverts and then goes to a piece of film. The film is exposed to light. Then when you develop the film, the thing that was *most* exposed to light is the darkest on the film. The thing that's *least* exposed is the lightest. So the darker will be the lighter part of the film; the lighter will be the darker part of the film, the more cloudy effect on the film. So then you put it in the enlarger, the enlarger puts light through the film. Where there's less darkness, less silver, the light comes in heavy; it burns through. Where it is thickest, the light comes through less and creates shadows and darkness.

You can think of a camera as: light comes in through a lens, goes to a film, then you take the film, flip it over into an enlarger or a printer, take the film, put light through there, and it goes through a lens and comes out onto a piece of paper. So we can think of an enlarger and a camera as exactly the same tool except that you reversed the source of light. The same thing is true when you're dealing with audio signal, but when you're dealing with broadcast radio signal, now you're talking different technology.

Radio waves go through the entire body, they permeate the body, and they affect the entire saltwater fluidic field that is your body, which is profoundly affected by radio waves and only minimally affected by audio waves.

Radio waves are going into cellular plasmas, and that's the difference between that and a binaural effect.

One of the things that's happening here is that there are corporations that have a vested interest in holding onto the notion that there's no interaction between electrical fields and the human biological computer.

The cell phone industry would fall apart if people really understood what was happening. Those vested interests are profiting by cell phones and microwave, power-line transmitters, broadcast television and radio stations where they're broadcasting waves into your body for money. A lot of that can be counteracted very simply by fighting the frequency with its own frequency, causing wave cancellation.

The sense of wonder is gone in science now. It's all a question of how can we bomb people with this? How can we kill with it? Will it actually fry people to death? How can we control people's minds? How can we control the money influx? How can we influence people to send more money to the government?

I'm not implying that this is related to how any individuals are functioning, but it has influenced the way that information is packaged and passed along. Information now is required to be passed in self-contained units that do not contain a little particle of *wonderon*, this little element; you won't find it on the periodic table, but *wonderon* is like: well . . . here's this piece of the puzzle and you'll notice as a piece of the puzzle it's got edges that make it relate to other stuff. We're ignoring the edges and focusing on the little piece of picture inside, and we're hiding from ourselves that this always implies something else, and there's more to the story.

So, in the photon-photon interaction question, it turns out that they do in fact interact and that can give rise to some very interesting theoretical speculations about the nature of a vacuum and the nature of photons, because if photons can interact with photons, once again it's understood that the interaction will be through some medium. If that thing is not going to be air, then it's got to be a vacuum. So how are they interacting? Well, that little piece of *wonderon* then leads a true theoretical physicist down another path of investigation, such as dark energy, dark matter, negative matter, negative energy, or something else.

If you're running a radio station and you have advertisers that are paying you money to put their promotion across the air, you have to ignore anything having to do with radio wave interactions because all they're interested in is how many listeners are actually being exposed to the message. That's all bottomline. They just want to know, "How many people are getting my message and is it worth the amount of money I'm spending?"

There's one binaural outfit that has sold 400,000 units of their binaural device. 400,000 of these things this year alone. Now they're probably worth a billion dollars today. We have no way of fighting those people, and I wouldn't fight them anyway. But they are going to fight us at some point and say no no no, we're right, and our thing works, and so on. I'm not saying don't buy their unit. What I'm saying is that it doesn't work to put sound in your ears and expect there to be a direct brainwave result. It's only affecting how you hear and how you hear might conceivably couple to how your brainwave eventually functions. But that would require some training and besides which, you don't need that; you can do it without training.

I made many different crystal radios in different sizes. I made quarter sizes, half-dollar sizes, dollar sizes and dime sizes, all working radios, but they also happen through wave-reduction and harmonic waves to be great Beta-Blockers as well.

And then I thought, how small can I get it? So I used my bat 64, which is the tiniest diode ever made and I made the smallest crystal radio ever — the size of a gold Maximilian coin from Mexico. I made it for fun, just to see if I could do it.

Most importantly, I made three variations in the quarter size, which can't be in silver; they've got to be in a nickel environment. I no longer use a green coil in these, so they're not sexy looking. I use a different circuitry.

We have a question from Jorge. He asks, "What would happen if a person quit the induction program after taking only a couple of Beacon trainings?"

I have had probably 20 dogs over the past 20 years. Every one of those dogs is well-behaved and likes it here and is friendly and is nice and has a good time here. They don't feel abused; they like it here. I follow the Barbara Woodhouse method of training. The one thing she says is never ever let the dog get away with anything, even once. Because once you do, you can never retrain that dog without a tremendous amount of pain. You'll need something like a shock collar or a choke lead, or something like that in order to train a dog once it has been allowed to abuse the training.

Believe it or not the same thing holds true of a teenager. Look, here's the parameters, now you're a teenager, you're entering a totally different realm here. You understand the consequences of some of these actions. So here's the parameters, okay? And here's what we're going to do. This is okay, that's okay, that's okay, that's okay, this is not okay, that's not okay . . . it's not that this is sometimes okay and sometimes not, this is not okay, that's not okay, that's not okay, that's not okay. If you break that discipline — and it is a discipline — even once, even once, it will get worse and worse and worse, and you'll try to get back on track again, it will become harder and harder and harder. So once you break discipline, you're asking for pain, serious pain to get back discipline again.

Now, that having been said, from the standpoint of training and discipline in the Work, I have the same exact training motto that the Seals use. “Failure is not an option.” That is what your motto has to be, because if you allow yourself to have failure as an option, *at some point you will take that option.*

Now this does not mean that you do not, from time to time, go into complete despair and wonder whether you'll ever be able to come through it. Discipline wouldn't be needed if it was all easy. Discipline is for those hard times when you feel like your ass is getting ground down to the ground, when you feel like you're going to scream it's so painful, when it's total torture to keep going. It isn't needed when the path is perfectly smooth and there are cookies all along the way.

This is not an easy path. It's not a smooth path. There are lots of obstacles, pitfalls, traps, wrong turns. That's where discipline comes in, and real discipline is based upon the motto: failure is not an option. Don't give yourself the option to fail, because if you do, you will.

What happens to people who start anything and then quit, is that eventually they find themselves in the same exact place sooner or later, but this time it's harder to get back to it and easier to quit because you've already quit once. The more you quit, the easier it is to quit, the weaker you become, the more strength you give to your opposing force, the Chronic. The more strength you give to it, the stronger it will become and the weaker you become.

Question from Henrik: “Someone on some sarcastic site commented on the black diode amulet in a blog — they commented on the black diode amulet design that the only problem with it is that you've shunted the primary phase coil. This inverts the field. Can you comment on this?”

Yes, I can. How qualified is the person who made that blog to comment on it? Have they bought one and tested it to see what the actual result is? This is an average person who's talking there on this blog, not a physicist talking, not even an electrical engineer because it's quite clear he doesn't know what he's saying.

When you don't care what the audio output is, it doesn't matter that you've shunted the coil, which in fact is not true — there's a second capacitor in series there. When all you're concerned about is the differentiation of two fields, it doesn't matter what the output is because you're not trying to operate it as a broadcast monitor, you're not trying to hear this as a radio. It's detecting radio waves and using them in a slightly different way than as a receiver. It's not an entertainment device.

He doesn't understand what he doesn't know. He doesn't appreciate how much he's not aware of. He's got a little piece of information from a high school course and that tells him that this inverts the phase, but what is the effect in the circuit of the phase being inverted?

He's obviously ill-informed because you have to combine four different sciences here; biofeedback, pioneer crystal technology and quantum mechanics, which is non-Newtonian and rather sketchy at best.

The fourth science is psychic. Something that physical scientists and bottom-line engineers do not understand is that there is actually a spiritual nature to all of this. When you start poking around in magnetism, electricity, heat and gravity, if you really are a physicist or a chemist, you understand it's all vapor — there's nothing really there . . . It's as if you were looking at a standing wave and saying, "Well, it's solid to me."

So there are psychic sciences, radio science and biofeedback technology, about which this guy knows zero.

I'm so much more impressed with the approach that MrfixitRick on YouTube is taking with this whole thing. He actually had an amulet on his test-bench, and plugged it into the equipment that he had available, and he's checking out . . . okay, what happens when you do this? And most recently he hooked into the circuitry and now he's playing around with touching the computer monitor, and he's finding that it has profound influence on the circuitry and has observed and logged what's happening.

Today you can find thousands of titles about the occult powers or properties of stones and gemstones, but in the 1960s nobody dared publish a book about it. It was one of those things you didn't talk about. Today, it's biofeedback, crystal radio technology and Parallel

Worlds that are considered "fringe" science.

Here's a weird one for you. I had a tarot deck called the Sun Deck. It was made by a friend of mine named Frankie Albano. He had his office right next to Los Angeles City College on Vermont Avenue. I used to pass by there every day and I'd stop in and we'd talk a bit. He had restored the Pamela Coleman Smith tarot deck as the Sun deck or the Albano-Waite deck. Prior to that it had been printed without the yellow in it — the colors were all lost. The Pamela Coleman Smith deck was superior to the existing badly copied decks, but the Lawrence Company printed it by changing the colors; I guess they figured they'd get away with copyright violation, which they did for years.

Frankie brought the Pamela Coleman Smith deck back. He gave me a set of the Albano deck — it was a first printing. He hadn't gotten the box printed yet, so I put it in a silk wrap which I like anyway for a tarot deck, and I put it in the glove compartment of the *Druidmobile*, which was a 1964 Ford Galaxie 500, which ran without a battery for three years. You tell me why. It started without a battery, ran without a battery. I don't expect you to believe that. I used to give free tarot readings at a place called The *Omnibus* which was across the street from Hollywood High School, around the corner from the Blue Grotto in Los Angeles, on Sunset Blvd., right next to the International House of Pancakes. That's where we used to have our group meetings.

I'd go over next door to the *Omnibus*; friends of mine operated it, but they had actually an *Omnibus* — a schoolbus — which used to travel with Further. We used to get up on top of the buses and do "unplugged" concerts in the park. We did Love-ins, Be-ins and Freak-outs in Griffith Park and Golden Gate Park with the Airplane and lots of other bands. I used to get rousted by the cops for doing *free tarot readings to people I knew in a private space* — a restaurant owned by a friend who'd given me permission to sit there and do this on Friday and Saturday nights.

I did one reading where my subject shuffled the deck . . . I did a standard cross layout, with four up on the right side. He didn't like the reading. It was coming up crossed Wands and physical obstructions, and so on. The final outcome was the ten of Wands. He did not like it, so I said, "Shuffle the deck again." He shuffled the deck again. The ten of Wands came out as the final outcome again. He was upset, so I said, "Let's try it again." So he shuffles the deck again and he hands me the deck, and I said, "I'm not going to touch the deck." I put the deck down on the table and said, "Why don't you do the layout. I'll tell you how." So he does the layout and I instructed him to cross the first card with the second and to put one on the bottom, one on the top, one on the left, one on the right, then down here, then one up here, then one up there — the way you do a normal tarot reading. So he did all that and it came out *ten of Wands*. And he said, "There's something wrong with this deck."

So I said, "Well, let's go through it and we'll lay all the cards out and we'll put them in suit and we'll see what's happening." "How many tens of Wands are there in here?" he says. We went through and we started sorting the deck out and there was no ten of Wands in the deck. I haven't told this to people, but you need to hear this because it'll help you understand what's happening with the Beta-Blockers.

I, of course, wasn't manipulating this deck. He had it in his hands all this time.

I got back out to my car at the end of the evening, about 2 a.m., and I opened the glove compartment to put the deck away. In the glove compartment is the ten of Wands sitting there face up! It had obviously fallen out of the silk when I pulled the deck out to go into the *Omnibus*. Now you tell me what's happening there. Lots of things like that have happened to me, and I know they've happened to you, as well.

So I happen to agree with the blogger that this thing is built wrong from a radio standpoint. He's absolutely correct, but he's ignored the psychic and biofeedback implications. He's only thinking about it inside the rigid box of electrical engineering; you cannot hook up a pair of headphones to the Beta-Block Amulet, and hear it. I agree, because that coil is taking the place of those headphones. He's absolutely right.

But he's the typical ultra-conservative scientist, because he's been *dismissive*. He hasn't actually checked it out to see something that might challenge his understanding and put him in the place of,

"Hmm, I wonder what's going on here?"

He's so excited about proving someone else wrong, and this is a little thing you should realize. Many people boost their low self-esteem and handle their fears by making others wrong. In the final analysis, he proved nothing except his own ignorance, and like many cowardly suicides, he's trying to take down with him as many as he can.

My answer is, as always, don't take anyone else's word for anything — try it for yourself and then decide what works for you!

Try it for yourself and then decide what works for you!

Crystal radio technology is at the core of Beta-Blocking.

Beta-Blockers ready for shipping.

Chapter Fifty

BETA-BLOCKER TECHNOLOGY
17 May 2009

A condenser is, in a small respect, a little bit like a battery. Basically, while the circuit is working, electrons are stuffed into one side of the capacitor. A capacitor will hold charge separation. Not an infinite amount of charge separation. That's a matter of how strong the capacitor is.

The fun thing about the induction and capacitor circuit, which they call LC circuits, is that they can be used to tune or select a given frequency. They do this by the interaction of the magnetic field that's produced by flowing electrons.

If you have electrons flowing in any conductor, electrons produce a magnetic field.

There's an experiment in the Cub Scouts. You take a nail, wrap wire around it, then hook that up to a battery so that you have a flow of current and that nail becomes what they call an electromagnet and you're able to pick up paper clips.

That is the basis of a metal detector actually, which we'll talk about later, about induction coils and how that works, but that's the basis of a 17th century metal detector, a field, which is tweaked by the presence of metals, field anomalies. Claude and I worked with the very early metal detectors. The Garrett and White detectors didn't actually work as well as the one which we later were forced to invent ourselves. But all that went into this strange mix of electronics that became the Beta-Block Technology.

There are some real fun websites that show you how to make your own capacitors for doing what they call total homebuilt crystal radios.

Home brew is the term they like to use for anything that you make out of spare parts.

Of the two plates in a capacitor or Leyden jar, what happens is, because of the electron flow, on one of the plates you'll have a surplus of minus charge and on the other plate you'll have a surplus of plus charge, which means the electrons tend to gravitate toward one of the plates and collect there. Well, actually they're attracted to each other, which is interesting from a theoretical viewpoint.

They have what we could call "electronic gravity effects". There's more "gravity" in one plate than there is in the other plate, more traction. Think of that in terms of charm.

Some of it is quite spooky in how it functions because what happens in an LC circuit is . . . L is inductance and C is capacitance.

As the electrons flow through the coil, it creates a magnetic field. Electrons will flow through a straight wire and produce a magnetic field at right angles. When you bend the straight wire into a coil it directs that field and causes that field to exist in a core. The formerly straight wire is now curved and placed close together, creating a magnetic field within it, and between each of these wires, which is called inductance, at a right angle to the coil — 90 degrees.

So in a straight wire, the magnetic field points out like a porcupine in every direction from the wire. In fact, that's one reason you can get 60-cycle hum out of any household wall.

It's really interesting to use that for some effects, and I actually use that in a field effect generator — if you put two wires together they will create a magnetic field, and if you have the electron flow in opposite directions they will attract and repulse each other at 60 cycles a second.

That "jiggle effect" can create a 60-cycle hum, and other kinds of hum as well. There are ways of adjusting those frequencies using what I call a magnet bomb, which is a target trigger system like a

Crookes tube.

Capacitor, condenser, Leyden jar . . . we used to make Leyden jars to make our own capacitors in laboratory. Dr. Hook had us make Leyden jars. They were actually an incredibly important component in the early experimenter's desktop because the Leyden jar could store electrostatic energy. It's called "electrostatic" because the energy comes about because of the *stored* situation. It's not because the electrons are moving back and forth that you get the energies, because basically in a Leyden jar you have a surplus of electrons in one plate of the jar.

In the Leyden jar, typically what you'll do is store the electrons there and then use them later for building potential in various devices. With the capacitor though, what you're doing is building up electrons that cause a back-flow such as the force from an incoming antenna signal.

Back EMF is like sharing an ice cream soda, and you could have two straws in there and with two people sucking on the same soda there will inevitably be a certain amount of backwash that gets into the soda from the other partner. Disgusting but true. This backwash effect is flowing actually backwards from the flow of incoming. It's backwashing and it does that kind of the same way as a hose full of pumping water, then you turn the hose off momentarily and it kind of backs up into the hose a bit and goes back into the pump.

If we start by looking at what's causing the flow in the first place, you have an antenna which is an electron pump . . . you have a copper wire. That copper wire is being acted upon by the radio waves. Now radio waves are both electrical and magnetic. They hit the antenna. However long that antenna is actually makes a difference as well.

Electromagnetic force is hitting the antenna, and that force is basically an arrow. And the arrow says "electrons go that way" and electrons go that way. Not all the electrons, but a certain number of them, go that way because the little electromagnetic arrow is pointing that way. So you end up with bulging electrons here because they went that direction. They got to the end of the antenna, but there was nowhere to go. So they got over there, and now they're sitting there . . . and they're piling up.

It's like in an auditorium. If you stand on the stage in an auditorium and you say, "Okay, everybody run to the left of the auditorium," people get up — not everybody, because you know how an audience is — most of the people run to the left, leaving empty seats on one side and more audience members over on the left. Then you say, "All right, everybody run back to the right." So then the crowd will run to the right. Most of them. Again, a few will stay in their seats and wonder what's going on.

Now this is why you get the selectivity because there's a natural rhythm, like on a seesaw. There's a rhythm to how fast a seesaw tends to go. The audience, because of this flowing back and forth, creates a rhythm. If the frequency of the beat of changing the signal saying, left, right, left, right, matches up with the audience's ability to flow back and forth, their own group enthusiasm, which is how you could look at back EMF, an electron group enthusiasm for flowing a particular direction; it will pick up and coincide and you will selectively enhance that frequency.

When you match that timing, you have beat frequency oscillation, back and forthness, which I use, I employ with all of my circuitry.

So that is in essence the relationship between a capacitor and an inductor. What will happen is if you want to change the frequency that's selected for by the circuit, you just increase the capacitance so that there are a few more members of the audience, or the electrons going from left to right, or you can change the inductance, the size of the arrows pointing the direction, and when you put the two of those together, it will give you your target frequency.

Selectivity simply means that you can actually decide the dominant frequency. It's not that all frequencies disappear except that one — they're all still there, but you can use a cursor and you can land on a particular point and say, "I just want to hear 760 kilohertz."

And here's where you get the real interesting effect. The way the frequency is selected for is that frequency is the one that is preferentially amplified because there's actually a little teeny amplifying circuit because we're saying we naturally want to flow at this frequency and the antenna is saying, yes, I've got this frequency also telling you to flow at that rate, they join together and it's a little bit like pumping your legs on a swing. It just gets higher and higher and higher, so the signal becomes more audible.

It's a magnification process that actually relies upon a composite or a cumulative effect. I had to decide which two frequencies were the most important frequencies and what my composite frequency would end up to be. I then put fixed capacitors in place that are, in effect, already tuned exactly where I want them to be.

When you want to generate ELF — Extra Low Frequency — you're working with ground waves rather than antenna airborne, and you've got to use a harmonic frequency because you can't possibly make an antenna long enough. I see ridiculous devices on the internet that advertise so-called "ELF" devices. It looks like something that came out of Buck Rogers or Flash Gordon, but it isn't actually going to do anything with the ELFs at all.

If you take a radio station, a nice 50,000 watt banger, and you turn up the volume really loud, and you put the mic next to a metronome — tick, tick, tick, tick, tick, tick, tick, tick — it's about 2 cycles. So you're in the 2 or 3 cycle range. You've got this little tick, tick, tick, tick. Remember this is AM — amplitude modulation — which means that you've got whatever the frequency is of your carrier wave, it's going like this: tick . . . tick . . . tick . . . so it just gets bigger when the tick is happening. It gets bigger in amplitude . . . the wave communicates that there's a louder sound there. So all they're doing is basically transmitting a *sound* — a purely audio signal — at what they're calling an ELF frequency.

In other words, they might as well do a binaural recording for the same effect because all that's doing is affecting the ear and the ear does not directly affect the nervous system. What I'm trying to do here is to work in the radio wave area where the radio waves are acting on a cellular level as well. They're acting on the water sack, on the fluid containment of the body. Every fluid containment is different. Electrolytically they're different because there are localized anomalies within each individual's electrolytic system. Just in terms of heat distribution and clustering effects between a heat source and a heat sink. Anytime you have a water flow, heat flow, electrical flow, magnetic flow, gravity flow, it behaves in the same general way.

If you apply the laws of flow to language, there's some interesting stuff that comes out of it. Shannon did a lot of work on information theory based on thermodynamics and came up with some very fascinating results. When you have a flow, you do not have a homogeneous even flow. Take a concert — we'll say a smallish concert, perhaps some 5,000 people in the audience, all contained in one area. That becomes the source. They open the doors at the end of the event, and now this audience has to walk down a hallway to exit the building, so now you have *source*, which is the auditorium where they're sitting, and the sink, or the destination, which is the exit onto the street.

As people go through the hallway, you know from your own experience they will not be a uniform, evenly-spaced homogeneous crowd. There's going to be clustering of various sorts all along the pathway from the auditorium to the exit doors. Now let's assume there's nobody that knows anybody in this concert. This is the lonely hearts club, and they haven't met anybody in five years. Since they don't know anybody, they're not clustering with friends to have a little chat. No, the clustering will happen naturally because of various factors. In information theory, or in what they call *negentropy*, there's a natural drive towards this organization because it actually *optimizes the overall flow*. Scientists would love it if uniform flow was more natural, because it's easier math. If every section was the exact same, then you wouldn't have to study the *whole* of the hallway. You could just study one little two-foot region and then understand the whole hallway from that small sample, which is the driving force behind most university-based science. Scientists want to study a tiny piece of the overall system, and then pretend that they have an understanding of the whole, but it just doesn't work that way. To really understand the whole, they would have to study the whole, because there's going to naturally be clustering and clustering does *not* occur evenly.

We all travel in our own way. Not everyone walks in a straight line, not everyone walks at the same pace, not everyone walks continuously. Some will stop and stare and look around and think, and so forth. There will be all kinds of reasons why there will be natural clusters forming in a variety of patterns all along the way through the hallway toward the exit point. We could call this hallway a *tunnel.*

So in this tunnel you have a tendency towards localized organization. Even marching troops will tend to cluster, although much more subtly than a shuffling unorganized mob of civilians.

Organization has to do with the interlinking of components. It has to do with, basically, order imposed on chaos.

Somewhere in the sixties negentropy and chaos theory really took on some life. There were people looking at it earlier, but in terms of research and acceptance into the scientific community, it wasn't until the late sixties and seventies that they really took off and gained momentum and some acceptance. Now you'll think, naturally, there will be tall people, short people, people with different strides so that they'll tend to cluster because of those factors. True, but that's not the point. Even if everyone was exactly similar, such as a fluid solution, because that's where these experiments were done, there will be pockets of cluster. Negentropy experiments were performed with a heat source on one end of a square petrie dish and a heat-sink on the other hand, where the solution was uniform — the same mixture of molecules. There are amazing experiments that can be done with this where you put a dye that is sensitive in proximity to itself and you can see the actual density of the molecules changing. They actually form into little swirls and eddies, like watching a tidepool.

I got into super-saturate solutions and mixtures of salts and electrolytic distributions and field anomalies within those. Solutions are a good place to look when you're looking at both mechanical force and electromagnetic force because electromagnetic force and mechanical force are related in the sense that they can both do work, but they do very different kinds of work.

When you have a heat source, what's happening is the molecules at the hot side are wiggling around faster. It's actually translational energy. They're just moving faster. Hot air moves faster than cold air. That's why certain sound and light phenomena may also occur.

We should talk about the invocation of Maxwell's Demons at this point because negentropy and Maxwell's Demons do relate even though they don't seem to. Maxwell discovered "demons" existing in fluidic and gaseous solutions. The demons have a particular dance, and even though you don't think that there should be any movement, there is. Not flow, but movement, zig-zag and swirl.

Science is not so much a study as it is a language. Physics is a particular type of language, chemistry is another type of language, mathematics is another type of language, and there are many other branches of science. Each branch of science is a different language if you think of it that way. It's a language and it's also, in a sense, a form of sculpture. What you're sculpting is your ability to visualize processes . . . and to anticipate and sometimes predict an outcome and a reason *why*.

There's a wonderful description of Tesla as a young man. When he built his first multiphase generator — motors and generators are the same, just backwards — he built it in his head. He just imagined it, he envisioned it. The thing that was definitely unusual is that he built this in his head and he set it running in his head, and then months later he took it apart and he made note of the portions, the parts, that were wearing down through use. This is a type of "thought experiment".

Einstein was able to develop his unified field theory and his general field theory and his special cases theory based on his visualization of what it would be like to try to run alongside a photon at the speed of light. He was answering the question, "If I were traveling at the speed of light and I turned on a flashlight, *would anything happen*?"

You can visualize things in a way that are predictive, in a way that allows you to look at the situation and answer questions such as, "All right, what if we made it hotter at one end of this heat bath, or made it colder at the other, or what if we had heat coming in here and in there, and then we put a little rubber duck over in this corner?" And all of a sudden you can *see the answers* to what that would do. All right, we don't use rubber ducks in science, so you'd reach on the shelf and pull down one of those black rubber stoppers out of an Erlenmeyer flask, but really, a little miniature rubber duck would have been the same, because you'd want to be looking at what happens if you put in a *perturbation*, excluding the fact that it's made of rubber.

Sculpting a model in your head allows you to predict and hopefully understand what's going on . . . and that's all through tricks. Because there are no actual little arrows in a radio wave, little arrows that go up and down saying "magnetic force, this way!" There's no little demon running around with a paint brush making an N on the north side of a magnet and S on the south side of a magnet, but I guarantee that almost all scientists, when they think about magnetics, have a little N painted on their heads over on the north of the magnet and a little S over on the south. All electrical engineers have a plus and a minus on their heads for the same reason. Or little tiny Es with a minus sign.

Predictability is a very important part of applied science, the science which serves the military-industrial machine. Some parts of science simply make observations. Denise's question was, "Can a Beta-Blocker be worn?" My answer was "I don't know. I don't know if I can make it that small." I now do know, and the Beta-Blocker amulets, rings and headbands are the result.

As I investigate I keep pushing and yanking things around to see what else can be done. Now I'm going back the other way to a much larger circuit with bigger coils, heavier duty materials, and also obviously more power in the operating system as well. I now know how small it can be made, so now the question is, *how powerful can it be*?

Here's a question from Jim and Mary Donald: "Is static electricity electromagnetic? Yesterday while wearing my amulet, every time I got out of my vehicle I got much stronger static electrical shocks than usual. I noticed, too, these never happen in the winter."

My answer is, when filling your gas tank at the pump, be sure to touch the car before grabbing the gas hose. Static shocks at the gas station can be explosive. Static electricity is definitely electromagnetic in that its movements will create electromagnetic fields.

Let's take your crowd again. You have a very excited crowd; they're actually panicky and they're looking for an exit from a theater. Suddenly you provide a path to the exit. They will instantly rush toward that exit. So when you have a lot of static electricity in the atmosphere built up or on the ground, on the earth, and you provide a path for it, the static electricity will leap toward the opportune exit . . . this is called “lightning”. The reason you see the flash is because it burns the air in its path. The burnt air creates a vacuum where the air was, then the air rushes in and you hear a clap of thunder.

There's another part to this . . . let me rephrase the question into a more generic form, which is, “While wearing my amulet, I noticed more of dot, dot, dot. Is the amulet causing there to be more . . .?” The answer could be yes, the amulet is somehow causing that, but I would look first to the pregnant Volkswagen syndrome. That's where either if you're pregnant for the first time or involved in a relationship which one of the members of the couple is pregnant for the first time, suddenly you start to notice preggies everywhere you go. And it's like, is there a population explosion going on? Well, no, your attention has been redirected and tuned. In the sixties, folks who bought a Volkswagen bus suddenly noticed there are a ton of other people buying the same vehicle.

If you bought a Chrysler New Yorker that had a purple paint job, you would suddenly notice that there is a sudden appearance of purple painted Chrysler New Yorkers just everywhere. They weren't there before, and now they're everywhere — every street has a purple maroon Chrysler New Yorker. It's the same year and everything, the same model, same make, same color . . . how can this be? It can be because your attention has been focused and redirected, and this can lead easily to a superstitious conclusion.

However, if you consider the situation of getting a new pair of glasses and suddenly noticing there are people and street signs and sidewalks and other vehicles, that could be a question of *being able to see*, so that would be an *amplitude* issue, or *sensitivity*.

As you wear an amulet your attention will be less diffused, less scattered, more focused, more directed, so it could easily be that you start to notice things that have been occurring naturally.

The body will tend to become more sensitive in certain psychic-intuitive areas, therefore, you will tend to notice things more, but you'd also tend to have an amplified experience or magnified experience because you're more sensitive, so your sensitization can be higher as well as your attention. You want both higher attention *and* higher sensitivity.

Sensitivity training . . . when we first started it it got bastardized so quickly by idiots at Esalen and places like that. We started sensitivity training in 1964. It was supposed to be a psychic-intuitive training in which your sensitivity became heightened to psychic-intuitive issues, so you became more sensitive to disembodied spirits, more sensitive to alternate realities, more sensitive to space changes where you actually sense yourself transitioning from one Parallel World or one reality to another, which you do all the time. That was the original sensitivity training, but then some people came along and said, "Let's make them more sensitive to relationship issues, to family, friends, and make them more able to cry more easily and feel more negative emotion." All this got completely tweaked. By 1967 sensitivity training had to do with relationships, not psychic-intuitive energies.

A group of my friends in 1964 got naked and went out to Box Canyon in Los Angeles, dropped a variety of psychoactive substances, and danced. They had a couple of bands with them. They called it "*Earth Day*" — they were celebrating *Gaia*. Now "*Earth Day*" is so tweaked that it's unrecognizable. That is true of absolutely everything; it all gets bastardized, changed and tweaked, so that no matter what the original intention was, at some point somebody turns it into a money racket. You turn it into a national holiday, then you have to go to Hallmark to get your card, you've got to send flowers so the floral industries are enhanced by that, you have to travel someplace so the travel industry is enhanced, you have to dress a certain way so the garment industry is enhanced, you have to have a certain spread at your table, and that means the produce and meat departments are enhanced, and so it goes . . . .

And of course there's a parade which means that the media industry is also enhanced — then, of course, in the parade you have to have a number of people who are guiding the parade, so law enforcement is enhanced, and everybody profits. There was a time back in the day when sensitivity training meant training toward the psychic-intuitive, toward the unseen nature, the hidden forces and the hidden side of the universe, the universe which is not perceived directly by the senses, the extrasensory universe of Cosmic Consciousness. That idea was completely, absolutely buried by the eternal quest for *money.*

In addition to the other aspects, you should not dismiss the possibility that you might in fact be subject to more static electrical shocks when wearing a Beta-Blocker just because, as you modify the electrical field of your human biological computer, its relationship with its environment will be greatly or minimally altered.

In other words, your electrolytic nature is bound to change as you wear these amulets, and they subtly alter the way your Chakras operate and interact.

Electrical engineers will typically ignore the psychic and intuitive areas. They're engineers for a good reason — they think in linear patterns. They can't think non-linear, they can't think outside the box. If they did, they would not be electrical engineers. Rigidity is part of establishment military-industrial science. The rigidity causes a reflex against anything outside the box of established science. So we're very fringy to electrical engineers.

Fringe just means, hey, you're on the edge, you're into an area that has not been well-trodden, it hasn't been written up in 1001 textbooks, hasn't been put in the university curriculum. Now, instead of saying "cutting edge", which sounds cool and groovy, you're characterized by reactionary conservative scientists as "fringe" if they want to discount or otherwise defame your inquiries and explorations into what they protect as their own private territory.

It's like saying, when you're interviewing somebody in law enforcement, you call them a *suspect. The suspect admitted under questioning that his name was John Doe*. It's a way of putting a slant, an emotionally loaded label on someone who triggers their territorial imperative.

There's also another explanation. Once you step into this fuzzy area of self-knowledge and the invocation of presence by the accumulation of parallel selves, you're under the obligation to actually start looking in your own environment, in your own situation. But what if you had previously been involved in mental agitations that were causing you to be clammy, non-dry? The drier you are, the more static electricity you will experience.

The more upset internalizations you indulge in, the wetter your skin will be. The more relaxed you are, the more your Ph will change and your galvanic skin response will change (GSR). If it spikes up and down, caused by internal perturbations, you might have waves of spikey electrical effects. I'm not talking here about midlife crisis, I'm not talking here about hot flashes. What I'm talking about is how your skin normally is — there's a certain pH — 5.6 — that's considered normal, but nobody has a 5.6 pH all the time. There's a certain galvanic level of wetness to the skin. Don't forget the skin is one single organ. It's an organ of the body. It happens that you live inside this organ, but it is an organ and it operates as a whole. It's one single organ like the liver, the kidney. The skin's galvanic state is profoundly affected by your internal state. When your internal state is good, you will have less sweat. When you're upset and agitated, you'll sweat more— not sweat like in normal natural sweat. It's normal to sweat when a temperature's hot. That's the way the body actually handles heat, but there are abnormal sweats, sweats that occur from emotional flushes. When that happens you'll also notice a particular odor, a very rank acidic odor called *adrenalin*. It's the same smell you will notice when you pass a stockyard where cattle are being slaughtered.

One of the reasons that steak tastes the way it does is because of the way that cattle are slaughtered.

Adrenalization is considered very good by people who are really into the body. People who are into the spirit do not consider adrenalin a good thing. So if you have more static electricity, it may mean that you're starting to adrenalize less and it also may mean that you're starting to spiritize rather than organicize. In other words, you're going away from the body orientation and more toward orientation in the centering and the spirit. This is a good thing.

So there are a lot of possible factors that enter into this, which we can test. It's in your province to take observation and look and verify for yourself what you think is happening.

In working with Beta-Blockers, one of the things I noticed is that when you're trying to integrate a new persona or a cosmic realization, there will be some galvanic skin response (GSR) when it's not integrated. You're just measuring the skin's change in resistance to electrical flow. When it's wetter it'll have less resistance, when it's dryer it'll have more resistance. That response is used as just one small part of the so-called "lie detector".

They'll put in electroencephalographic information, electrocardiographic information, galvanic skin response and a variety of other responses that all go into a composite picture to indicate that there was an upset, or confusion, or conflict about a question or statement. In fact, the word "conflict" is the key word in terms of the lie detector, the polygraph test. If you just took one tiny section out of the polygraph, which means many graphs, it wouldn't give you a true picture. It's just one bit of information that is unrelated to other information.

So if you just took the information that came off of the skin detectors, all you're going to get is, was there more sweat or less sweat? Was the body more conductive or less conductive in relation to that question or statement? And by the way, it's never instant. There's always some lag, so you actually have to measure the lag repeatedly. There's a whole series of things you do in the beginning of a lie detector test to determine whether someone is actually "in the test" or "not in the test". In other words, what was their time factor — how much of a lag there is. Essentially There will be a lag in the electrocardiogram and the encephalogram, and the galvanic skin response. Those three in particular have a definite lag because it takes time for the reaction to set in. The time can be in microseconds, but there *is a lag*.

The lag will vary depending on what's being measured because they'll measure the chest volume. Those little springs that they have around somebody's chest when they're in a lie detector that's measuring expansion and contraction of the chest, and that'll have a different lag from the skin response.

They're looking for things like not just a sudden inbreath, or gasp factor, they're also looking for suppression of those things. Anything that looks like a suppression of information coming to the lie detector is important to the person reading the lie detector test.

They also look for long-range patterns. They'll look at the breathing rate overall and that influences how they interpret the other data.

Hyperventilating will have a great deal of influence on galvanic skin response, for instance. In addition to that they're looking at myoelectrical impulses, meaning, are there muscle twitches happening? And if so, where are the muscle twitches, and are they common? Do they occur all the time or do they occur only in relation to certain questions? Are the muscle twitches being suppressed to hide something? . . . and so forth.

So, all that having been said, galvanic skin response still does have a useful effect in terms of *integration.* When there is a non-integration of some new experience, there will be an agitation of galvanic skin response, which will occur as long as that agitation is in effect; and as soon as the agitation is gone, which means the integration has completed, the galvanic skin response will settle down into a steady state.

That is generally what happens with everything that has to do with change. The body resists change. This is important to understand. The body seeks *homeostasis*; it wants things to not change. So as long as things are in a seemingly steady state, the body is happy. When any change occurs, the body becomes unhappy and immediately starts to invoke the Chronic, which resists the change. We can disarm the Chronic by use of the Beta-Block, allowing the Essential Self to equalize and justify and harmonize and integrate, until everything's in a more or less steady state again.

One thing you can look for when you've just integrated a new parallel universe persona (PUP) is how long it takes for the integration to occur; typically with each individual it will vary and for each individual persona it will vary as well. Some are harder to integrate, some are easier to integrate. Some are more intrusive, others are very passive.

How would you like to remember *every detail* of a parallel universe life? What a confusion that would be on this life! So you don't really want to have every single memory forever! For the integration process to work really well, you want it so that the Essential Self's currently manifested persona wins every time, the one that's in the home world universe, the one that is directly and most vitally attached to the Essential Self. The Essential Self has to win every single time.

What happens is when a persona is pulled in, it tends to dominate even over the Essential Self, temporarily. When the integration occurs, the Essential Self reasserts its dominance. So you have to revisit the question of machine or Essential Self, not once, but over and over again, many, many times. Every time you bring in a new persona, you're going to have to actually actively and consciously integrate that persona into the composite manifestation of the Essential Self. You're trying to build a cluster around the Essential Self that is fully justified, equalized, harmonized. Just as each parallel universe persona is x-number of Kakus away, when it's pulled in it has to be harmonized.

You're pulling in a parallel universe persona that requires integration and part of that process may bring up issues that either you've never had to deal with before or you deal with again in that you're not revisiting old issues for this persona, you're visiting new issues coming in.

Do the amulets actually bring in parallel universe persona? The answer is yes they do, but slowly and gently. The Beacon pulls them in, and wearing a Beta-Blocker amulet helps keep them calm while the integration process occurs. Basically it causes less struggle in the integration process. So that's one use of the amulet in the parallel universe procedure.

Anytime you bring in a new persona it's a stranger in the tribe, and the stranger has to go through all of the ritual of integration in the tribe. So you're actually building the Tribal Self.

We have to establish that the tribal self is an internal tribe. It's kind of like the Portuguese Man-of-War, you become a colony in which the whole is greater than the sum of its parts.

Can the cumulative effect of the Beta-Blockers in my being affect those around me?

Yes, indeed, they can . . . in two definite ways — one is, as your electrical force changes, and the other, as your behavior changes. As a result of both of those, people around you will be affected by the change. They'll say, "You know, something about you has *changed*. I don't know what it is, but it's really nice, whatever that change is, it's really good. I *like* that change, whatever's happening to you." Somebody will walk up to you and say, "What are you into? I want to be into that too, whatever it is, whatever is going on with you, I like it."

Others will only feel the effect of the Beta-Blocker and they will have a mild effect like a *contact high*. So both of those effects will occur — effect with the Beta-Blocker, but also an effect with the Beta-Blockee, meaning you. So your change of being will affect those around you profoundly. Sometimes for the good, sometimes they won't like it because it's a threat to their Beta lives, their territorial imperative. They may not like you *changed*. They might prefer to have you be totally predictable, completely static, whatever it is you were *before*. *The organic self always resists change*.

So as you change, people around you will react one of three ways: it's called the one-third, one-third, one-third principle. One-third of the people will love the change, one-third of the people will be totally indifferent to the change, and one-third will hate the change. Be happy with that — one out of every three people will be happy that you are changing, one out of every three will be totally indifferent to your change, won't care what's going on with you, and won't understand how to deal with it or what to do with it and won't care, and then one-third will *really* resent the fact that something about you has changed, and they can't pigeon-hole you anymore, keep you in the same category. They have to re-categorize you.

Question from MrFixitRick. Can an electrical device be brought to life?

Ah, the Pinocchio effect, I call it. I mentioned Billy Whitefoot. James Redfeather and Billy Whitefoot worked in Cosmo Street. Billy Whitefoot was a shaman of the Apache. Billy was at Cosmo Street and every so often he would give a lecture along with Ivan Caldas who was a Dervish of the Qadiri Rifai sect and he would talk about everything in the universe being alive — everything.

The question is detectability of life. Can you detect that life? Can you actually respond to it and detect it? And also, does it have a consciousness that can detect you? So your definition of alive has to be really clear here. But from that standpoint, I think I know what you mean. Can it be brought to life? The answer is absolutely it can be brought to life. And that's not restricted to just electronic devices. John Lilly and I used to give lectures quite often, and one of those we gave at the ICC.

We talked about *silicon lifeforms*. What you're familiar with is carbon-based lifeforms — human beings, little furry animals, fish and fowl and mammals and reptiles and amphibians, carbon-based lifeforms. But did you ever stop to consider that a computer is actually a silicon-based lifeform? To itself, it's very much alive and it may regard humans as *not* alive because by its definition of life, carbon-based lifeforms would not be included.

An electrical plasma is technically the same as a biological plasma.

The thing actually that comes to mind more on this topic is where are you? If you're in what you think is an extremely solid "bang your head into it Newtonian, George W. Bush universe," then *nothing* is alive. In fact, *you're* not alive in that universe. It's all just mechanics. But as you shift into other worlds or perceptions, a truer perception of the world, suddenly the stuffed animals on the shelf say "Hi, how you doing? Where you been? What's going on?"

They learn to fly and attend your readings, meetings and dance and drum circles. If you consider yourself an electrical field plasma, then yes, everything is alive. If you consider yourself to be a physical, organic body, composed of mechanical forces, then nothing is alive, including *you.* In a strictly mechanical Newtonian universe, you're only as if alive.

The more you can see that little electrical plasma buzz that's happening in the environment, the more life you will naturally appreciate in your immediate vicinity. Then there's the question of how much respect, how much Beingness do you give to these various items?

That is one of the barriers we hit when we're 3 to 5 years old, when you learn to keep your observations and knowingness about the universe to yourself.

If you choose to disobey that social imperative and to even mention that you're noticing things outside the "box", you'll be sent to someone whose job it is to stuff you back inside the box and keep you in there. When you're noticing things outside the box, you're considered *dysfunctional*, and dysfunctionality can't be allowed in a society, because the society would then lose its control over you, and societies that have no control are *chaotic*. An antichaotic society is a society which exerts its consensus reality on all its constituents, and if you dare even mention anything outside that consensus reality, you will be punished. You'll receive an electrical prod to get you to go back behind that fence like the other sheep.

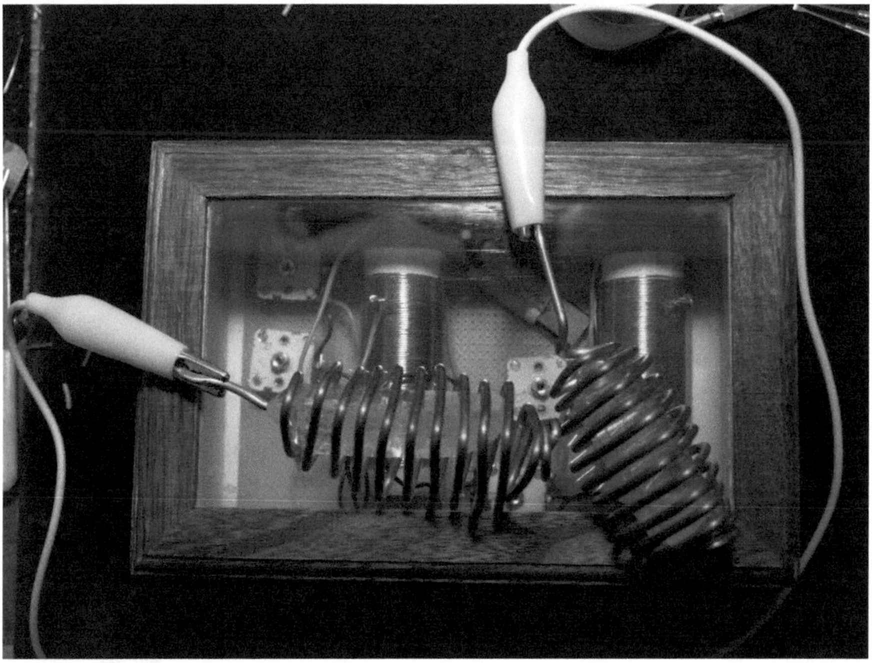

Beacon Automapper from above showing crystal grips and double coil construction.

Claude leads discussion about the Chronic at Beta-Blocker Workshop, 2009.

Group discussion on the chronic, Beta-Blocker Workshop, 2009.

Chapter Fifty-one

THE CHRONIC AND THE BETA-BLOCKER
May 24, 2009

If you're still plagued by the Chronic in all of this, you're going to be in for some heap of trouble when you start messing around with parallel universes, so I would say, grit your teeth and bear it because you're going to be in for a lot of reaction. If your body reacts badly to being out of ice cream, or to the soccer game being canceled, or to not being able to get the particular fashionable clothing that you were lusting after this past several weeks, or you go on vacation and you freak out because the traffic is too slow on the highway . . . if those things cause you to freak out and panic, imagine what a parallel universe is going to do to you. And that freak out and panic is the Chronic. That's what that is. Or the whimpering, bitching, griping, groaning, complaining, whoosing out, the compulsion to take center stage and not only to have a problem, but to be the problem. This is the realm of the Chronic, and if you are still there, if you're still demanding attention of others because you personally don't feel all that good, I guess maybe you're in for some serious deep trouble.

You're probably not going to read T. E. Lawrence's *Seven Pillars of Wisdom*. You'll probably watch the movie though, *Lawrence of Arabia*, in wide screen. In one memorable scene a subaltern is standing around with Lawrence and Lawrence is holding a match between his thumb and forefinger. He holds the match until it burns down to his fingertips, and the match goes out by itself in his fingers. He does nothing to put the match out. He just stands there and holds it. The subaltern tries it himself, and he lets it burn down to his fingers and then he lets it drop. And he says, "Ouch, that bloody hurts! What's the secret?" And Peter O'Toole as T.E. Lawrence says, "The secret is . . . not minding that it hurts."

Well, you're dealing with the Chronic. The secret here is also *not minding that it hurts*. Let the Chronic scream and cry and whimper and bitch and gripe and groan . . . let it carry on. You must have to have the will to go past its little territorial tantrum.

I'll tell you the characteristics of the Chronic. It says, "Well, I don't quite understand. I have a million more questions. And if you answer those I've got a million more. You'll never get past my questions, you'll never, ever get past my barrage of questions because my questions are my defense, my doubts, my worries, my criticisms." That's what the Chronic puts up to defend itself in its *flight or fight* combat, its battle against reality.

The defense mechanism and the Chronic — those are the things that actually are preventing you from being five years old. That's the thing that's preventing you from actually exploring other universes. That's the thing that keeps you here, keeps you in Beta — and Beta is actually what it's protecting — that's the precious, that's the ring of power — the Beta is the ring of power. That's what the precious is all about. That's the ring of power. That's the thing that all of you is invested in. Do you understand? If you destroy that you'll destroy yourself . . . my God, you can't do that! But that self — that particular self — is called the "ego". And that is the precious, the ego is the precious. There's a real self and there's the ego. The ego is the thing you protect, the real self, you can't even see because the self is so much in front of it.

Your mind is your biggest enemy. It's a trickster, and a Beta mind is the biggest trickster of all. It will convince you of all kinds of things that are absolutely not true, just to get you afraid. You know you love to be afraid. That's why horror films are out there. People love to scare themselves, to tickle themselves with fear.

What is your soul mate? There are some folks whom you travel with from lifetime to lifetime; you always end up with pretty much the same dungeon party, the same party of folks. You always travel with the same group.

In any given human lifetime, you'll get into a series of relationships, usually it's serial relationships. *Each one of which* is with *your soul mate*. You're convinced at the time that this is your soul mate, then you discover that the individual that you're with is a total creep, and you bust loose out of that, and then you get into another relationship. Now *that's* your soul mate for sure. And it isn't that they're *not* your soul mates, it's that what happens is that things change, people change. So you start out with your soul mate, and then you drift into a different reality, and in that reality, the person you're with isn't the same person you were with a few months or years ago. They look the same, they suggest the same individual, but they're not the same. So you might find yourself with your soul mate, but then something else ends up in that same place. If you're looking for your soul mate, you can use a Beta-Blocker to attract it. And the reason that'll happen is because as you "turn on" as an essential self, as you turn on and your manifestation magnetism is stronger, your magnetic center is stronger — those who are naturally attracted to you will be even more attracted to you, so it's easier to find your soul mate, or soul mates, your fellow travelers. But, the secret is staying with them, and it isn't always *what they look like*, it's *who they are*.

It's hard to wrap your mind around this. You're trying to use a Newtonian or Einsteinian mind to wrap yourself around something which is actually far into the realm of quantum mechanics. There is no direct experiential data in the non-Newtonian universe that you can draw upon — you can't see it, you can't feel it, you can't taste it, you can't touch it, but here's how you can experience this:

You may be traveling with someone who looks the same year after year, but if you look at them very closely you'll see that it might not be the same individual that started out in that body. So one day you'll wake up and there will be a stranger in your bed and it'll be the person you married. And you'll wonder *how could that be*? How could that happen? How could they change so much?

They didn't change, because what you call a person is actually a *place*, a focal point, a locus for consciousness, and then consciousness may move on. If you don't know how to move with that, you'll end up in a space where that individual is no longer there.

Mostly people get in trouble in a relationship because relationships *are* trouble. Actually, *there is no right answer* — relationships *always* lead to trouble. Either you can handle the trouble or you can't, but there's no such thing as a relationship that doesn't, at one point or another, get into trouble. There's a good reason for that, and that's because two or more people are involved in any relationship and they have separate and divergent interests, separate and divergent beliefs. You try to find somebody who has the same beliefs, the same interests, takes the same political positions, you have a lot of things in common.

So you go to harmony.com and you fill out this form and you tell them all the things, and then they try to find somebody who's compatible with you. Compatibility is the number one issue in relationships. If you really are mature, you don't *need* to be compatible, you can *let somebody else be what they are*, and who they are, and what they're doing and still be with them, and enjoy them for what they are, rather than try to make them a mirror image of what *you* are.

In a relationship the two partners are typically trying to make the other partner align with them, their interests, and their desires, and their internal mental commands, and their realities, and so forth, their preferences, and on and on. In point of fact, if you're in a real relationship — I mean by that an objective relationship which transcends the self — you're not going to demand of your partner that he or she conform to your reality.

Every choice-point has at least two paths, two divergent paths, so there are realities where you were never born, where your brother or sister or spouse was never born, where neither of you were ever born. Then there's somewhere that both of you are still together, there are others where you have died, there are others where they have died, there are others where you were born, but you never met, and then there are others where you were born, met, married and separated. Then there are others where the other just went away. Every time there was a choice-point, another reality was generated in an infinite wealth of alternate worlds. I know that's hard to accept, but that is the way it is. There are actually an infinite number of alternate realities involving *you.*

And I'd like to speak to the idea of infinite and nearly infinite, almost infinite because there's a big difference there, and you need to know what that difference is.

Let's say you have little shot glasses and the shot glasses have just a whisper of Vermouth, the teeniest, tiniest amount of Vermouth. If you have a large number of these shot glasses, then we add them altogether, it may add up to be a gallon of Vermouth. But if you have an infinite number of shot glasses they will add up to be an infinite number of gallons of Vermouth. Now if the shot glasses were full right to the top, and you still had that large number of the shot glasses, then it might add up to be a hundred thousand gallons. But if it was an infinite number of shot glasses, it doesn't matter, they still add up to be the same infinite amount of Vermouth. It doesn't matter how much Vermouth you have in the shot glass, if there are an infinite number of them, it'll always end up being an infinite amount of Vermouth.

So you can actually make a nice little equation that says, infinite number of shot glasses with one drop equals infinite number of shot glasses full to the top, and it looks like it's, "Well, wait a minute, how can that be?" Well, that infinite number is a very different thing. We tend to think of infinity as just really large. It is not just really large; it is so beyond really large that a teeny tiny single atom of Vermouth in that shot glass is the same as a full shot glass when you add it up over the infinite number of shot glasses.

So infinity times the smallest amount imaginable is still infinity, just as zero times the largest amount imaginable is still zero.

As one begins to embrace parallel universe travel, suddenly you have no home. You're eternally homeless, you have no place to rest your head. I call myself Gorebagg the Lost. I always assume that I'm lost; I always assume I have no idea where I am, and there are no proofs. If you're looking for proofs, for certainty, this is the wrong set of universes to do that because you're actually in a sea of uncertainty in an ocean of improbability. So when you're looking for anything that is solid, an anchor point that you can really solid ly grasp, no matter how solid it seems at any given moment, it can dissolve in your grasp.

When Morgan passed away, people asked, "What happened? She was at a school; how come she died? Well, she did not die, she's currently in India. She didn't end up being reborn in India, she went to India to work with and study with someone there, but it looks like she died here, to those who were watching this process. She's actually quite alive and quite well in India. Not in this particular universe, but she is in a parallel universe in western India, near Mumbai, living with a family, cooking for them, and it's easy to contact her.

In most human cultures you're made wrong if you don't *miss* somebody that has apparently passed on. You're supposed to mourn them, and if you can't mourn them, if you haven't got time to, you *pay* mourners to mourn for the dead. The Eygptians, Sumerians, Babylonians, ancient Greeks all did this.

You can use the Beta-Blocker and its characteristics to overcome your Chronic's natural fear of success and allow yourself to become successful and prosperous, but you have a *good reason* to become prosperous.

You're going to find yourself living in spaces where you're starting to notice that the people around you are more peaceful and more harmonious. You are gradually moving from universes that are dangerous to universes that are more peaceful and harmonious. As you work with these Beta-Blockers you will gravitate more and more toward the more relaxed universes, the universes that have more peace and harmony. That's the whole secret to Buddhism is that you don't change yourself, *you change your world by changing worlds*. Not by changing the world itself, but by switching universes. You can't just go from here to there; there's a particular route, a rather circuitous route, not always a straight line. The straight line is not always the shortest distance between two points.

Of course, giving you tools to accomplish this means that I have to consciously and knowingly enter the dangerous universes you have fallen into so I can introduce those universe-switching tools. It's a risk I knowingly undertake — pun intended.

UFOs are visitors from Parallel Worlds using telepathically controlled devices to transverse the membranes!.

UFO visitors are not from outer space. They have visited us throughout the ages, and we can visit them back!

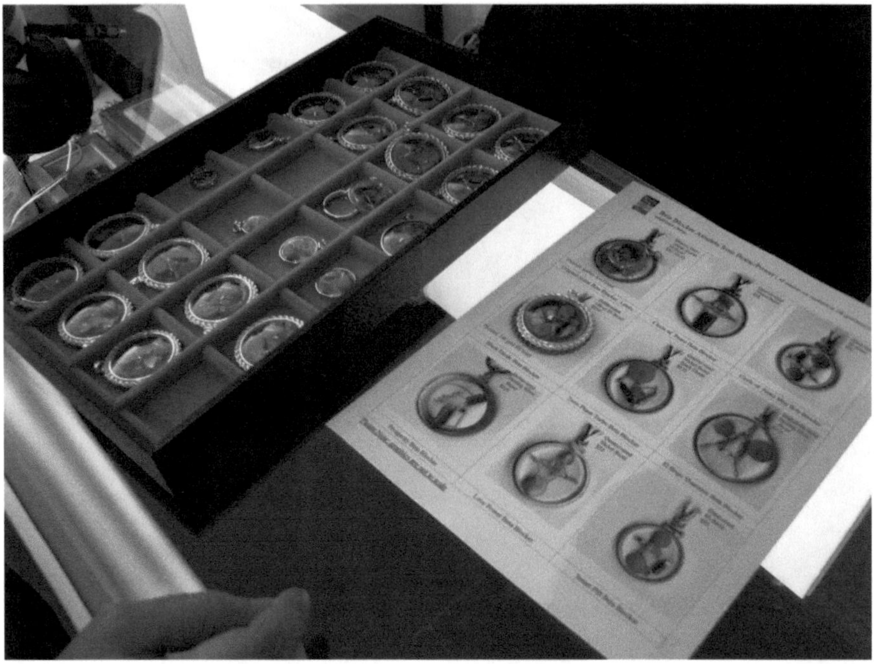

Beta-Blocker amulets made by E.J.

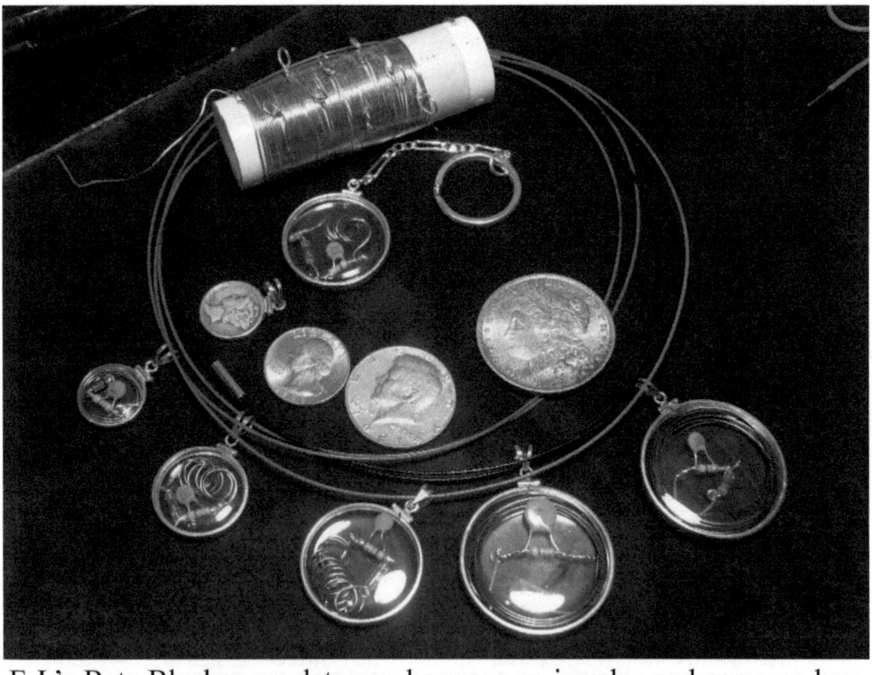

E.J.'s Beta-Blocker amulets can be worn as jewelry and some as key-rings and money-clips. Rings are also available.

Chapter Fifty-two

ELECTRONIC MOJO

One element of the Beta-Block Amulet is commonly overlooked by some, and observed and acknowledged by others . . .

It is the "Mojo" of the amulet, in which we consider the amulet as a magical or spiritual apparatus, in much the same way that we would consider a crucifix, star of David, Chai, Mezuzah, Corno . . . or, indeed, any amuletic jewelry worn by all peoples of all cultures around the world.

Far and above the scientific principles of radionics — which are very much fringe science and not demonstrable by any known means other than that radio waves exist and that they do indeed affect the skin, bones, muscles, nervous system, brainwaves, thought patterns and emotional states of all types of biological life including plants — Beta-Blocking amulets have another mysterious qualitative, unmeasurable "something" about them.

You cannot prove by any scientific method known the efficacy of Holy Water, the Crucifix or Saints' Relics such as the ashes of ancient Tibetan Holy Men or the Knuckle Bone of St. Peter . . . but they are associated with miracles of all sorts, and rightly so.

Millions have bathed in the Ganges and gone to Lourdes . . . no one questions the Miracle Power of the Shroud of Turin even though it has been proven to be no older than the Middle Ages, and literally billions of amulets, charms and Holy Relics are carried by billions of people around the globe.

Amulets, charms and other Sacred Objects are made in a variety of ways, but all of them carry in them something far greater and more powerful than mere scientifically-accepted "Mojo" or magic.

The Beta-Blocker Amulets I produce have a fringe science background and are dependent upon radio waves, which exist throughout Earth's atmosphere at all times, permeating everything on the planet with the full spectrum of radionics, but there is something more to them than just the radio wave functions, something indefinable and unmeasurable . . .

I call it "Electronic Mojo".

I cannot intellectually and with scientific math prove my thesis that the slight amount of radionics penetrates easily into the 11th dimension at a distance of less than 10^{-27} millimeter, which is what I call "One Kaku" of distance, and that it opens pathways into the branes of parallel universes which you can travel, but with a little practice with my inductions, you can prove this to yourself.

Someday I hope that we will see laboratory "proof" of this effect, but in the meantime, the evidence will have to be very subjective, like the evidence for life after death, near death experiences, astral projection and creative visualization, among other spiritual effects and practices.

Blind Faith? Not necessary! Just practice, practice, practice.

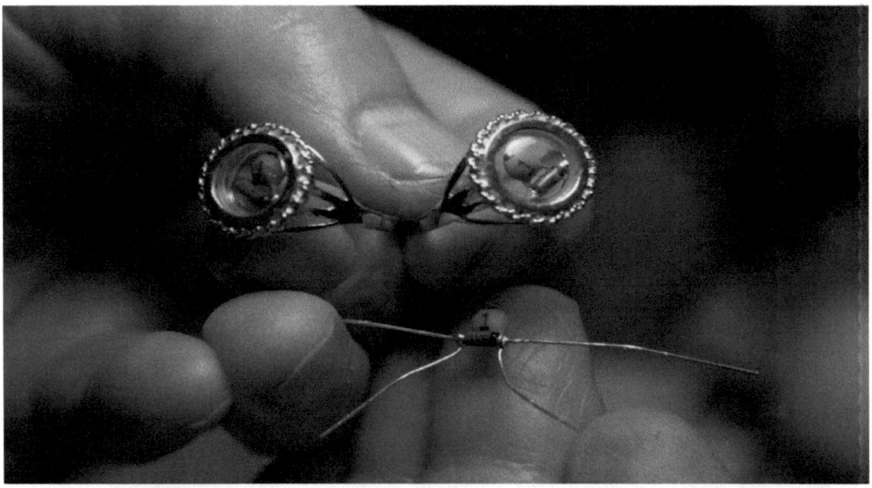

14k gold Beta-Blocker rings by E.J. contain miniature crystal radios!

Chapter Fifty-three

STAR STATIC

The digital revolution has left as its victim radiation from the event that signals the first moment of the universe — The Big Bang. An analog radio set that has been specially tuned between broadcast channels receives static, the result of energy left over from the super-hot primordial universe as it existed in the first few seconds of creation. This static is known as the Cosmic Microwave Background Radiation, and its discovery in the 1960s proved conclusively the Big Bang Theory and Parallel Universes, which are today all the rage on the Science Channel and talk shows . . . but shamans knew about this ages ago. Science is just now catching up and is slowly awakening to the real nature of the universe, which is far more plastic and spooky than anyone ever thought.

AM radio signals are very different from FM and television signals — AM or Amplitude Modulation is capable of registering faint "background" radio signals broadcast from a variety of sources, among which are the formation of Population III stars just after the Big Bang, Black Holes and clouds of Dark Matter and Dark Energy, invisible by any other means.

How this energy can be harnessed for Inter-Universal Astral Travel is something which we are now exploring with the Beta-Blocker AM Crystal Radio Technology, which has proven itself to be a very powerful tool for the shaman of the 21st century.

One certain way to harness this mysterious energy from beyond our galaxy at the very beginning of Time and Space is to combine it with the most powerful known psychic energy available to the shaman, which is generated as a by-product of sex energies.

Simply by wearing a Tantric Star Beta-Blocker on a headband or sweatband will turn on an internal psychic trigger that harnesses Stellar Static from the Big Bang, Black Holes and other cosmic sources, but then it must be directed in order to use it as a psychic shamanic tool, and that's where the Tantric Star Induction comes in handy.

The Tantric Star Induction guides the user to the correct harness of Cosmic Background Radiation which comes to us as static. By correctly tuning the Tantric Star with a combinant frequency adjustment obtained through the use of twin-tuned capacitance miniature circuitry in the Tantric Star, we can trigger the subtle psychic energy state of the 6th and 7th Chakras situated at the Pineal Gland and the Parietal Lobe respectively.

Om Ma Ni Pad Mi Hum . . . Hail the Jewel in the Lotus — the thousand-petaled flower situated at the crown of the cranium. Here we find the highest chakra, which processes and originates pure thought and consciousness and which can be cluttered with information through the imposition of the Beta-Brain.

This "Sahasrara Chakra" has its infinite petals of spirit reaching into the infinite cosmos. In Sanskrit, the word "thousand" equals "infinity" and as a matter of proven fact, in the mind there are no limits.

The primary nature of the Crown Sahasrara Chakra is the Essential Self, an unmeasurable entity of non-identified selfness which is just one manifestation of the Unified Field of Consciousness which is commonly referred to as Cosmic Consciousness.

It is the Crown Chakra that is our main connection to Cosmic Consciousness. We pass our will of manifestation through the Crown Chakra into the lower Activity-Chakras to manifest the Cosmic Will, and it is through the Crown Chakra that we enter and leave a material body.

A human brain with its billions of matrix-connected nerve cells makes more connections per microsecond than the number of atoms in the universe, reaching and affecting more than 100 million sensory receptors in the body, and many trillion synapses in the nervous system.

This means that the non-Beta mind is millions of times more sensitive to the Multi-Verse consisting of an infinite number of universes and environments than our external sensing organs.

Blocking the Beta-Mind and going within the Crown Chakra gives us a method to access the 11th Dimension, which has no locality in time and space. Time and Space are functions of the local universe, and are not present in the Mega-Universe beyond the Big Bang and its results, which are Stellar Matter, Dark Matter, Dark and Light Energy and Atoms, all of which produce the Gravity Wells which in turn produce the illusion of Time and Space.

Each Chakra gives us a contact point with a number of dimensions of smaller waveforms and faster vibration of frequencies which have their manifestation in radio waves detectible as AM static.

Through the triggering or activation of the Crown Chakra by the act of Beta-Blocking, we can instantly contact the 11th Dimension, the central axial dimension which has infinite speed beyond the speed of photons and zero wavelength, which by definition exists everywhere and nowhere, creating the effect of Spooky Action At a Distance, noted by Einstein and other physicists dealing with Quantum effects.

The 7th Chakra is your gateway to "cosmic consciousness", and provides you with a direct way of perceiving the highest reality, that of the Mega-Universe, which can be said to be a patterned set of order within an infinite chaotic field of undulating energy.

Consciousness is a way of realizing or recognizing the Self within this chaotic field, and differs from the ordinary search of Stanley and Livingstone by the fact that the seeker is also that which is sought.

The Beta-Mind is the organization of experience and information which forms an electronic matrix of ordered data which is triggered by association and environmental provocation. We spend our years sorting information about the environment and trying to make some sense of it while constantly adjusting and adapting ourselves to the demands of the world so we can keep our job, maintain our relationships and stay out of the rubber room at the funny farm.

It is this continual attempt to bring order out of chaos that causes confusion and a brutal and corrupt separation from our true state, that of Infinite and Eternal Cosmic Consciousness.

By harnessing both the radionic energies of the Big Bang and those of the seven Chakras within but not part of the human body, we can reattain the Waking State of Cosmic Consciousness and resume our rightful path through Space and Time to our ultimate destination, the Here & Now.

Our ultimate destination is the Here and Now . . .

Chapter Fifty-four

MULTIPLE DIMENSIONS

I want to talk about superstition, parallel universes, super gravity, String Theory, M-theory, the 11th dimension Riemannian space theory in relation to M-theory because I think it's very, very relevant. I think Riemann had it pinned fifty years before anybody else did.

Even if it's right in front of your face, it's not necessarily true that a primate human being is likely to actually see and comprehend something even if they do see it, and parallel universes are not actually seeable. By their very nature they're parallel; they are not actually in this universe. If they were in this universe then they would be part of this universe. They're extra-universal universes, membranes that exist outside this universe.

When you're contemplating how this all works, I want you just to think about an 11th dimension which actually permeates but does not touch this universe or any other universe. Every one of those universes is, in some way, connected to the 11th dimension, but isn't actually part of it, and they're not part of each other. They all exist in the same space at the same time, but because they're at different frequencies, they don't actually come into contact except occasionally, and that's what I want to talk about.

Rippling and undulating membranes — we know they undulate. How do we know about these things? Well, there's one indicator that tells us that there are Parallel Worlds, and that one thing is gravity. All other forces are very powerful. Gravity is the weakest, and in fact, it's so weak it almost doesn't exist. You might think otherwise because gravity holds you onto the earth. At the same time, you can lift objects with your bare hands. And if gravity was as strong as let's say, electricity or magnetism or heat, you wouldn't be able to lift anything. It is now popular wisdom among physicists that gravity actually is seeping into our universe from somewhere else.

They have magnets these days that come with warnings — they are so powerful that if you hold the magnets and push them together so they'll attract — they never come apart — you can crush your fingers between them. Planet Earth is huge compared to a walnut-sized supermagnet. If Earth was made of the same magnetic material as those little magnets, we would be attracting stuff from all over the asteroids.

The composite mass of the galaxy would be attracting other galaxies at a rate that would cause the expansion of the universe to actually become a contraction of the universe. The whole thing would just implode on itself. Just to give you an idea of the weakness of gravity . . . if you jumped off a 6-foot ladder, you could feel the strength of gravity in your shins when you hit the ground, but relative to all other forces such as electricity and magnetism, gravity is so weak as to be ignorable. In fact, if it wasn't for being stuck on the planet, we would discount gravity — we would say it's too small to really worry about, let's leave it out of our equations completely. You could actually leave gravitic factors out of most physics.

Gravity is not actually a force in this universe; it's a force in a nearby universe which is seeping into this universe. If you assume that gravity exists within this universe, most physics equations don't work, but as soon as you assume that gravity is seeping into this universe from elsewhere, suddenly equations — especially cosmological and quantum equations — fall into place and start working.

In physics one of the guiding lights is elegance. When you're defining a mathematical expression, elegance has a lot to do with your sense of correctness because there is really nothing to prove that math is correct. Experimentation will verify the mathematics of physicists, but there's really no such thing as proof. It's just a question of verification. There's always an array — a bingy-bunch of equations — that could, once you plug the numbers in, give rise to similar experimental results. For example, drop an object and it falls at 32 feet per second per second. Now you can get that number by plugging values into different systems. And they could be equally correct in that all of them predict the same experimental results so that when the experiment is done, it says, "Ah, 32 feet per second per second, therefore, Professor Smith's equation is correct." When physicists are sitting around speculating, it's not just the fact that you're able to properly predict experimental results that counts — it's also how cool and groovy or elegant your equations happen to be, which means that your math is a "neat fit". It's like carpentry where you have a dovetail. And the dovetail can be sloppy and careless and crude, or it can be elegant and neat. An example of an elegant equation is $e=mc^2$.

If you include bathroom stalls as a publishing venue, it's been published more than any other equation you care to mention.

In order for you to have an equation that incorporates gravity, instead of $e=mc^2$, you have a page-and-a-half of math that gives you the gravitational components. So when somebody comes along and says, "You know what guys? If you just imagine that gravity is coming from another universe, then you could account for gravity using a much smaller, more elegant expression."

When we work in equations we don't work on paper, we work on blackboards because you can erase whole sections of things and replace and refine and adjust them. One of the easiest ways to handle questions of the 11th dimension has to do with getting gravity out of the picture and putting it where it belongs, which is in a nearby membrane.

We had lots of disagreement between the super-gravity and string theorists, which was resolved eventually by placing gravity outside our Home Universe. The 11th dimension is approximately 10^{-27} millimeter across and an infinite length. So it's really, really narrow, but infinitely long.

Horace, my dad, once said he was going to buy a spaghetti farm 3,000 miles long and a half inch wide, and he had a clever way of doing the harvest — you simply roll the spaghetti up from one end.

What scientific proof breaks down to is predictive ability. If you try to look at a Newtonian universe, it stays together. It'll fall apart if you start looking at extremely distant stars, or if you're looking at gravity wells as opposed to orbital structures, but from a practical point of view, Newtonian physics works most of the time.

Before we had high-energy particle physics equipment — accelerators and magnetic bottles to hold plasmas — as you could use simple photography to do relativistic experiments, which is how they first verified relativity — but if you weren't doing that exact photographic experiment, all of our collective experience was perfectly well-predicted using Newtonian physics. There was no need before quantum theory to stretch into new territory. If you traveled back in time to present String Theory in the1800s, you'll be put in a funny farm by even the most advanced and tolerant physicists.

It's only been recently that we've had the ability to even start looking in non-Newtonian directions. You cannot today examine any question of physics without stumbling at some point into the mechanics of quanta. Quantum physics has at its very foundation M-theory and the theory of multiplicity of dimensions — the multiverse, it's called now, which is a plethora of universes.

Universes are now considered to be membranes in M-theory. I happen to subscribe to M-theory simply because I have seen them. I've seen membranous universes, universes that are donut-shaped, torus-shaped, wiggling and vibrating and undulating. I've seen them from outside; I know what they look like. They're self-luminous and any so-called mystic, any voyager — astral voyager — will tell you that these universes look like sort of wiggly donuts and squiggly other things as well, that are interlocking membranous things that are constantly undulating and that are rolling and some that are like waves crashing in on each other.

So, what was it that caused the Big Bang in the first place? Shamans can tell you what caused the Big Bang. Two membranes crashed against each other. They were close enough in frequency that they actually interacted with each other. When you look up in the sky and you see the Milky Way, that's seeing your galaxy edge-on, but it's actually a spiral. The Milky Way galaxy is starting to collide with the Andromeda galaxy, and when it does, the two galaxies will intermix with each other and the black holes may or may not merge in the center of them, but nothing will actually touch.

There's so much space between everything that the galaxies themselves will actually pass through each other almost as if they're incandescent gases, but they will affect each other in terms of roil and turbulence and disturbance and they'll form lumpiness. Lumpiness is part of any sustained system that is fluidic or gaseous in nature, determined by heat and fluid dynamics. Those lumps are very important in understanding how the universe works. Undulation, ripple and jiggle are very important in understanding how all systems of interlocking universes actually interconnect and interact with each other without actually being on the same frequencies and without actually touching each other. You can verify this when working with magnetic fields which don't actually touch each other, but they do interfere with and co-regulate each other.

The 11th dimension is permeated by a multiplicity, an infinite number of dimensions.

How do we know that we have parallel lives in other universes? If even a very small fraction of those universes contain parallel lives, and it's an infinite number, then guess what, folks? It's an infinite number of parallel lives, because a partial infinite is an infinite.

Let's say only one percent of infinite parallel universes have any form of life, and one percent of those have intelligent life, multiply that by infinity and you're once again back to infinite. That's the amazing thing about “infinite”. You multiply anything other than zero by infinity, and you come up with infinity. So if intelligent life exists on Earth, multiplied by infinity, then there's an infinite number of Earths much like the one you're on now.

There's a theory that Claude and I are working on called Divergence Theory. Divergence Theory has to do with branching at choice points. We have made many games to illustrate choice-points.

The thing about the Divergence Theory is it may have less to do with the causative factors of the actual creation of universes and more to do with the stringing together, or assemblage, of the sum totality of your linear experience as a Being, of those universes. That is more cogent by far. It's called "Thread of Consciousness".

Let's say that you find yourself in some science fiction universe and in that universe you happen to be in a video store with a lot of videos to watch and maybe a lot of time available to watch them, but it's an infinite number of videos, so you're not going to be able to watch all the videos. So there will be some pattern that you use, even if it's just randomity, to determine which videos you decide to watch.

Darwin called that "Natural Selection". This just happens to be how things are working now, but they're not the only way they could work. There could have been other options. You could have decided to pick and exist in only those universes in which all of the lifeforms had the same morphology, but based on our experience that's not really a major criterion, because you'll find yourself going into very different morphological lifeforms. There is a thread, there's something that makes them similar. And this has to do with a higher order of functionality.

If you find yourself at a choice point, and there are, let's say, for simplicity, two possible choices — either you do this or you do that. Apparently it's more significant to us to find alternate universes that offer choice than to find alternate universes with the same morphology. The continuity of flow seems to be more important than making sure that you have six fingers on each hand and ten toes on each foot. I think that's right. Apparently morphology and topological consistency are not that important to us as Beings.

Shapeshifting is secondary to a higher order of choice. There are some individuals who select their manifested lifeforms based on what they're going to look like. Blessed are they, they should be known as Clairol blonds. I don't mean that to be genderized because it's not. There are some folks who choose by how much money they will have, or how much power, or influence or by how many friends they will have in that lifetime to text with.

But then there are others who choose in a much higher order, more along the line of a spiritual nature, a work orientation. Their work-path is more important to them than what they look like, or how much money they have, or where they live, or what they drive, or how many friends they have that they can text with, or how much sex they have with good-looking, sexy partners.

What I wanted to get at here is that the multiplicity is unthinkably enormous. As a matter of fact you can't just use your Beta-Brain to think your way into an understanding of the infinity of multiplicity of parallel dimensions. Like every learning process, it requires experiential data, experience personally experienced by you. I therefore have developed a series of levels of training related to experiential experiences in and with parallel universes.

The first level of training is simply to get your feet wet, just to introduce you to the idea of parallel universes and get you connected up and give you a little more weight in terms of collective selfness. The collected self is what you need to start with in Level 2. There is a "you" in many Parallel Worlds. We're only talking about strapping to the ones that are most functional for you to get to Level 2. At Level 2 we're going to start jaunting from universe to universe, moving around through and between Membranes.

In the "Just Because Club" exercises, there was one exercise that we didn't know ahead of time, but we discovered in hindsight that it was a filter, and that was an exercise which, on the face of it, seemed simple enough. You go to a museum and you spend time in the museum — an hour — just being in the museum without ever once looking at the exhibits.

I produced a book called the *American Museum of Natural History*, which is a photo book that consists of photographs of the entire museum without a single exhibit being shown.

So how hard could this be? A significant number of folks managed to attract the attention of the guards and either got thrown out of the museum, or in some cases, put into an observation ward.

The reason that people get picked up for psychiatric observation is because they behave very oddly. Not looking at an exhibit does not require bizarre behavior, but for some reason some people triggered into bizarre behavior in order to fulfill this "Just Because Club" concept.

The vehicular translation method of space changing I leave for an advanced course. In the meantime, Level 1 now consists of 30 CDs, Level 2 consists of one CD, Level 3 one CD, Level 4 one CD, Level

5 one CD, Level 6 one CD and Level 7 actually is where we check to make sure that everything has been running right and that nothing got left undone or got done that shouldn't have been and so forth. So Level 7 is a checker level. I will be using for that a device called the GSR-2, which is now currently available at your local pharmacy and on eBay and Amazon — 500,000 units of this particular device are out there presently. I bought two dozen sets which I will make available to people in the checker context at Level 7. Level 8 is preparatory to the next set of levels.

I have the idea in mind that someday someone is going to try to stop us from exploring these domains and from conveying information about this material because it does, in fact, have military applications.

I can tell you that because I was involved in two of the military applications at the very beginning of 1962 to 1968, I was involved in military applications of this particular science, meaning that the United States government and the Soviet government, at that time, were actively involved in the exploration and inquiry of Parallel Worlds and alternate dimensions, so it's not something that we're making up, it's not something that people are just now figuring out. There actually were agencies that involved themselves in extra-dimensional, extra-universal exploration and military applications thereof. I cannot tell you those things. I can simply tell you that those agencies and activities existed.

Many science fiction and fantasy writers, horror, detective and western writers are writing from the standpoint of peering into alternate dimensions, Parallel Worlds, and other universes in order to derive their story ideas.

In 1976 I was talking with Frank Herbert. I fortunately managed to record the conversation between the two of us, and in that conversation I said, "You know, Frank, I saw Arrakis when I was shamanically dreamwalking." He said, "Well, that's actually how I came across it. I was shamanically dreamwalking, lucid dreaming, and I simply used it for my story of *Dune*. The only difference between you and me is, I beat you to publication." He said many, many people that he talked to had seen the Atreides family, and all he did was report the story as he understood it from the point of view of the multiplicity of viewpoints that he was able to look at it from.

Ted Sturgeon said the same thing to me many times, that he was able to travel into other dimensions and from those dimensions he was able to extract — he called it the Prometheus effect — stories from other dimensions — parallel universes as we call them now.

Evelyn E. Smith, the great fantasy writer, talked very often about morphing and shapeshifting, as she traveled from one universe to another, and often referred to herself as a living dragon under enchantment into hominid form.

We're taught to use the Beta-Brain to discount all this as fantasy. I have often said that *reality is a crutch for those who cannot face fantasy*.

Well, fundamentally we have no clue what is really going on. The 11th dimension is so close to every point in our universe that it actually is an unmeasurable distance away — which is why my devices work, because they only have to travel 10^{-27} of a millimeter or less to get where they're going.

Artist Odd Nerdrum certainly is looking into other universes. Salvador Dali is probably the most infamous artist who used the Prometheus Effect to grab images from other dimensions and bring them into view on a canvas. Giacometti also had that same visionary ability. So did Joan Miro. Marc Chagall I'm not so sure about, whether he was looking at an alternative universe, or he just didn't know how to paint.

I'll tell you a little bit about scientific superstition. You can often misattribute something. You have an observable phenomenon, or set of phenomena, and you can decide why that's happening, theorize as to why it's happening. The "why" is always subject to superstition. In other words, you may be thinking that there's a cause and effect lineage there somewhere that you can actually see and predict, but it isn't necessarily true that what you have decided is actually the reason why something occurs.

Here's one for you: there has not been a single day in recorded history in which there wasn't a war somewhere on the planet. That's an interesting factoid that's easily verified. Some wars lasted thirty years such as the Thirty Years' War. Some lasted a hundred years.

War of the Roses is a notorious example of a long, long war — the Hundred Years' War, the Thirty Years' War. The Trojan War actually lasted about a decade-and-a-half.

Most people don't realize that the 18th dynasty, which was a rejection of the priests of Amen Ra, and the replacement of those priests by the priests of Akhetaten, only lasted 18 years. One of the interesting things about the 18th dynasty is that the city of Akhetaten, known today as Amarna, only existed for 18 years, so you can see a little time frame, only 18 years long. It had been buried and forgotten about. The entire 18th dynasty, as a matter of fact, didn't exist until Carter rediscovered the 18th dynasty by discovering the tomb of Tutankhamen, the so-called boy king.

I don't study these things. I simply know about them. I know about them because I've been there and done that. I can tell you on a day-to-day basis what daily life was like in the 18th dynasty. I can tell you what daily life was like at various times in Ancient Rome as well, as can any shapeshifter time traveler such as myself. The difference between us is I don't have a memory block, and if you do have, then you're going to want to use the Beacon to take down that memory block so that you will actually have full memory of everything you've ever done in your Long-Life. Time is an illusion, lunchtime doubly so. Time is not linear. You enter and leave time frames the same way that you would enter and leave a room. I therefore can travel to the 1800s after I've been in the 2100s, in the 2000s and the 3700s, and so forth. It's simply a matter of moving through time the way you move through space, and that will be the subject of Level 9, which actually will be the first intermediate level. You'll be time traveling, and you will actually travel through time.

I know that there are financial considerations, but at some point you should have your own Beacon, your own headphones, your own Walkman CD player and your own grippers. The reason for that is because a lot of what you'll be doing will be related to work that you'll be doing on your own. You may not be able to come to a center, but you do need to do that work with the Beacon.

We have Beacons placed in 26 countries at this time and I think that's pretty doggone good. I would just as soon give them away, but I'm not in a position to do that. I'd like to make the Beacon accessible to virtually anyone who wants one.

We have a question from Peter Heineman: "Since we are constantly exposed to various electromagnetic waves from devices of all kinds in our environment, do we need to worry about these random effects interfering with the effects of the Beta-Blockers? If so how do we get the maximum benefit from the Beta-Blockers?"

The Beta-Blocker has a limited blockage of those interfering broadcasts. There are what are called intentional radio broadcasts, and then there are natural radio broadcasts, such as those coming from the stars. Stars are radio broadcasters; they actually are radios. They're like giant Tesla Coils. There are gas clouds and all kinds of things that are generating massive amounts of radionics. And as you know you're exposed to microwaves of all kinds coming from everywhere. And cell phones work because you're being bombarded by signal, no matter where you are, except for dead zones, you're being bombarded by signals which are microwave signals. They're signals coming to and from the cell phones. Those are satellite broadcasts. So those are intentional.

There are also relay broadcasts coming to you from places like mountaintops which are microwave antenna banks, whole banks. There are also transformers, there are electrical transformers everywhere, and those are producing a massive amount of cancer-causing electrical field. There are also minor things, such as in your car. There is a generator in your car, which is actually a Tesla Coil. It has somewhat shielded you by a steel body unless you're driving around in a fiberglass body or plastic body car, in which case, you're not getting shielded, but then again, you're in front of a computer every day, and the computer is generating all kinds of things.

The oscillator in there, just the time, the clock oscillator alone is generating enough frequency to really mess up your nervous system. So a wearable Beta-Blocker is indicated. And that is one of the things we use it for, but on top of that, when you're working with your Beacon, you want to cut out many of those things. However, sometimes they're of actual use to you although they're harmful to the body. They can be useful to your voyaging.

So, we don't cut out everything. We only cut out those things that are harmful to our voyaging and to our explorations. We don't really care too much about the deleterious effects of radiation because we're being bombarded by so much radiation in our culture today, so much radiation.

The 60-cycle hum that you get from your electricity . . . when your electric goes off, you notice how quiet it is . . . it's deathly quiet! It's as quiet as the grave. Until the electric goes off, with nothing else on, nothing else operating, just the electricity being there, you're getting 60-cycle hum. And so when you're out in the desert, you're not getting that, but what you are getting is a kind of a song; it's a very high-pitched sound. That's coming from the stars, directly from the stars. It's called Star Song. And every desert dwelling shaman knows what Star Song is: it's called the music of the spheres or celestial music. It's also called the Schuman Harmonic or the World Harmonic.

Now that exactly is what we tune to, 7.8 hertz. When you get that harmonic it shuts out in many other things that are harmonic to it, such as 1530 megahertz is not good, 1530 kilohertz is not good, 730 kilohertz is very, very bad for you, and above 1400 to around 2700 kilohertz is very harmful. Not in itself, not that frequency. It has harmonics — wave harmonics — that are buzz harmonics. Anything that produces a sawtooth wave or a square wave is not going to be very friendly to your body.

There are a lot of waveforms that are not good, not friendly to the body. There are also waveforms that are not friendly to the spirit, let's say, and you can feel those. People will vibe you. When you're in a mall and you're walking past people, you feel the vibes coming off of people. Now they are connected to Parallel Worlds, and you're getting that vibration from a multiplicity of dimensions. It's not just them. They're not transmitting it by themselves, they have help transmitting it, in the sense that they're getting what's called auxiliary feedback superheterodyne oscillation, which is oscillation amplified by other dimensions.

Remember how close everything is. Alternate dimensions are less than a 20,000th of a micromillimeter away. They're very, very tiny distances between universes — 10^{-27} millimeter is a very tiny amount of room between you and it. It's right in your face.

These things are freakin' close to us, and they require almost nothing in the way of energy, that's to say, amperage or voltage, or any other -age that you care to name to transfer influences to us. They require very little micro Henrys, millimicro Henrys, milimicro pico quasi semi pseudo Henrys. There's very little energy required to make a synaptic leap from one universe to another because all of them pass through the 11th dimension, which is actually as close as you can get without being in the same space at the same time.

The traversing of these dimensions is very easy. That's why it's so easy to be shifting in and out of these, because there is this membranous undulation which occurs. That undulation causes you to be constantly shifting realities, constantly changing realities. So let me give you Darth Vader's answer when he detects the presence of Obi-Wan Kenobi aboard the death star. He knows he's there someplace. He says, "I detect a presence I have not felt for many years. Obi-Wan Kenobi is here." His companion says, "Quickly, let us stop him before he can escape." Darth Vader replies, "Escape is not his plan."

So, Peter, I hope that answers your question. Escape is not my plan, shouldn't be yours — don't worry about escape.

When I was working with linear accelerators in the 1950s, and we built a thing called the perhapsatron, which was a very early cyclotron, we didn't know about shielding. So I was told that whatever genetic damage was going to happen to me had already been done, which is interesting because my family has been sterile for seven generations.

"All these other electromagnetic forces, do they affect the Beta-Blockers?" asked Yanesh.

Of course it affects the Beta-Blockers. Radio waves permeate everything that you live and breathe and work in. The atmosphere is bombarded; it's actually full of radio waves, television waves, cosmic rays — tons and tons of radiation is hitting you at all times. When I'm building my Beta-Blockers, I take into account that those things are always there. The only thing I don't have to take into account is gravity because gravity, for all practical purposes, does not exist. It technically is of such a minuscule amount of energy that it actually has no effect whatever, but radio waves do have a pro found effect. In fact, I'll tell you how you can measure that effect.

There are some people who receive radio signals in the fillings of their teeth. You've heard of this, I know. It happens when you make fillings of the same material that we use in our crystal radio. The human head is a crystal radio because you have salt water and acids. The acids create electrolytic effects, so you get a radio operating in your teeth.

There are tons of bombardments, and in fact, there are also some vicious intentionally broadcast bombardments coming to you from Russia and China, and all of my Beta-Blockers — especially my wearable Beta-Blockers — take that into account that you're getting the woodpecker from Russia and similar effects from China and North Korea. Originally the woodpecker was meant as a detection device at ICBM, but it has since become a weapon in the hands of the Russians and the Chinese, and they're still using those today even after the Cold War.

Civil servants in Germany all wear Beta-Blockers, and they cost the German government a fortune, about $2000 for a wearable Beta-Blocker. They're issued these to prevent the Russian and Chinese woodpeckers from interfering with their brainwaves. It's a very real thing. This is not something that we're just making up. You can check this out and see how much radio wave warfare is going on out there.

The Beta-Blockers are a counter-measure to radionic attack. They're also a counter-measure to psychic attack and they help immensely in psychic attack issues. Psychic attack has two main causes: one of them is intentional and the other one is accidental. There's the intentional psychic attack — someone hates your guts and vibes that anger, that hatred, at you. That's what monkeys do all the time — but you don't want to be a part of it, and you don't want to respond to it; you just want to block it.

Unconscious psychic attacks happen when somebody hates life, is disgusted with their life in particular, hates everyone who might have a nice life, or at least a life that looks less painful to them, and generally fumes. You'll find these people in every mall. These are the people who usually are handing out the cheese-whiz on crackers. They poison the atmosphere. There are people sitting in the work-station cubicle next to you at your job whom you can feel. You go to the water cooler just to get away from them for a few minutes. A Beta-Blocker will block the psychic attack that they unconsciously direct nowhere in particular. In most psychic attacks, you're actually not the target. You just happen to be in the way.

It also is a very good effect against radionic attack — attack of unconscious radionics coming at you from a variety of radio, television and microwave sources. Does it protect you against everything? No, it doesn't, but it gives you an alternative path. Using the combination of the Beacon and black diode, you can actually weave a path for yourself that avoids most conflict. You'll never find a path that avoids all conflict. Many Beta-Blocker users have written in saying, "I know that when I was wearing my black diode, I avoided a very bad automobile accident," or "I was going to get into a fight with my spouse and I didn't, and I think it's because of the Beta-Blocker."

It does help you to weave a path between conflicts, skillfully avoiding them like dancing between the raindrops.

Magic Skunk is a powerful ally against psychic attacks.

Beta-Blocker users, 2009 workshop.

Beat-Blocker users at the Neal St. Center, 2009.

Chapter Fifty-five

Beta-Block Experiences by Beta-Blocker Users

To me, just an ordinary person, the most basic stuff going on about the Beta-Blockers is creativity, God, It, the Being . . . that's the bottom line.

So when you're wearing it, even having thoughts about it, gets you into that space of Creation, the potency of it, the beauty of it, the exploration, it's the door, you're opening up that mysterious place where you can contribute to the Universe. It's a reminder, a support that you can go there. It's a reminder of Creation itself. Some invisible hand is touching you saying Yes . . . the Beta is all about Nos and the Beta-Blocker is all about Yes.

B. K. — Penn Valley, Ca.

On Saturday the 11th I went to bed at the group house. E.J. had given us a Super Delta Inductor, which was designed by him to induce or help induce a person into the Dreaming. In my dream, I was with Ricardo in a room with tables and chairs. The tables looked more like tables used for students than for eating. They were brown, and there was a window somewhere, perhaps behind Ricardo, perhaps to my right, or perhaps it was of the kind that is only detectable near the limits of one's peripheral vision — you know, the kind that is only there when it is not being looked at. In this room there might have been others too. I remember I was in constant contact with Koyote and we were having a conversation, standing up, him leaning against the edge of one of the tables, his arms might have been crossed. He might have been wearing his hat, he was probably dressed in white.

My dream was about four hours long as far as I can tell. I remember entering the Dreaming almost immediately after falling asleep. Sometimes there is the definite experience of time passing before a dream-state is fallen into, but this time there was no time lapse at all, which is very interesting, because E.J. might have said something about the Dreaming happening as if we never fell asleep. Nice. Waking-up was also immediate.

The sense of time was perceived then, and I "knew" upon waking-up that about four hours had gone by since I first laid down on my bed. In rationality, it had been a little over four hours since I had gone to bed. I was fully charged, energized, as if I had had a long six to eight hours total rest and not a beat was lost. It is amazing, not that E.J. told us that was exactly what was going to happen on the night we went to sleep with the Super Delta Inducer, but that it actually happens — not to mention that it is probably the first time it has happened to me without so much memory loss before this day. Koyote confirmed my experience of the Dreaming by relating a statement that he had also been there, naming the day and the time in which we had both experienced this meeting in our dream.

Regarding my second day I would like to add some more experiences I had. The manifestations of the effects of being not in beta continued throughout the morning. It seemed to me that, having had a very interesting and deeply restful dreaming night the night before Sunday, I was still confronted with a body that wanted to fall asleep very soon into the workshop while E.J. was talking. I attribute this to the body and the mind's habit of falling asleep before entering the non-beta states.

I remember a very potent experience in particular. I opened my eyes and the fist thing that grabbed my attention was the stare of a little guy in a white t-shirt sitting at the bench in front of the keyboard which was behind E.J. and Claude. His wild red hair was cool looking, and his stare was actually very intense. I immediately noticed the presence behind the eyes. Everything else around me was background, even E.J.. This lasted for about one or two seconds. Then I was back in the room with everyone participating in the workshop. I remember feeling grateful to have experienced this being, for I know they live, and E.J. and Pat are two of the ones that most show it because they actually talk to them. That is cool. I attributed that experience to the Beta-Blocker's working effect on us. Of course, the fact that I'm being exposed to cleansing radiations in the presence of E.J., and specially during these types of workshops, presents some difficulty when attempting to discern how much influence the Beta-Blocker has had. I heard that the HGA manifests herself in so many ways, and it is not unusual at all that phenomena such as the appearance of beings normally unseen to the naked eye become available in such a place (the Institute) and within such a presence (the Teacher).

It has been a custom for me to marvel and take joy in the vistas that are presented every time we are blessed with the opportunity to work in the garden — the vegetable garden for me in particular-where I have been exposed to other dimensions and other times. I have, for instance, many a time been struck frozen in a particular spot within the physical plane of the garden — a seemingly insignificant spot accessible through any ordinary means of detection by any of the locals within the human dimension — where, after a day and more of relentless radiation exposure within the safe confines of E.J. workshops, dimensional windows and doors appear before my very eyes, showing vistas of rolling hills which seem to expand into a radiant light and away into an endless space. The beings have mostly been in human dress, usually they have that invisible shine, and they are working on the land of good earth and grass and plant and rock, caressed by the wind and bathed by the invisible sun. They are of a different time, and they are the pure workers. When I come back to my body, I see that the beings I am working with and the land is surely accessible to others as well, and that in some way we are transposed within many worlds and dimensions that do not exist in time and space; moreover, I marvel at the idea that such a place has been reached through this machine I inhabit and wonder at the thought that eternity is happening there.

Then again, there are the other beings that walk or glide and fly through the tree branches and the leaves out in the other gardens and within the magical halls of the house and the barn — shapeless, non-human, shadowy . . .

When has the beta state not been breached and neutralized through other means? The drawing exercises which take me into spaces of quiet and stillness, the silent participation in the emptying of a storage facility with twenty others, a demonstration of an organism that is usually, under ordinary means, scattered by many different personalities, each with ego, believing themselves to be the only ones, not realizing that they are but a part of something bigger. What a genius demonstration! By the simple and effective introduction of the technique of using silence, we end-up with a being composed of us all; a being with a centralized ego, silent, no-beta. Nice! What a powerful tool to introduce to us! We take that insight back home, to Riverside, and the wish to apply it with and introduce it to our group is born.

Now we have a tool just as effective as the Beta-Blocker, but more useful indeed. That's the powerful benefit of such a body as the Institute, that such a simple living example, in the form of an exercise, is introduced to our little group, and we are given the opportunity to try it out in living expression so that we may carry the experience in our bodies and infect the rest of our body, the group, The Tequihua Foundation.

Carlos — Riverside, Ca.

I wear my radio near my heart where it works well with my heart neurons. The entrainment that occurs with my heart neurons and the radio is more harmonious than wearing the radio near my throat. My throat reacts uncomfortably to the radio's presence.

I have many years of practice using the HearthMath Institute techniques — Freeze Frame, Heart Lock In, and Quick Coherence — all of which entrain my heart rate, breathing rate, and sine wave activity. I use many techniques that help me consciously shift the Beta processes to managing movement so I can function more in a more fully present state of being.

When I first put on the radio — I had visual and auditory responses from my guides that were very positive. The closest I can come to explaining the reaction my guides had is to reference you to the story line of Marion Zimmer Bradley's work with the Darkover series. In her first book of the series, Landfall, she wrote about the interaction of the humans with the native beings of the planet. Those beings lived primarily in the trees and possesed psychic capabilities. They were tall and slender with slight blue tinge to their skin. They had long straight white hair and were hermaphrodites. Their movements were lithe and flowing, tone of voice was high-pitched and ethereal. They danced and sang all the time. That is all background to explain the experience I had when I first put on the radio. My guides began dancing and singing with the other entities (some of which had the graceful appearance of the race of beings described in the books about Darkover) in the space. It was very harmonic and joyful.

Although I do not have this same experience every time I put on the radio, I am able to access that experience with ease when I put on the radio. Wearing the radio functions as a tool that assists me in remaining present and providing an opportunity for the Absolute to have an experience using my machine.

I noticed too that I am able to move a little more easily into the state of presence when not wearing the radio.

V. S. — Austin, Tx.

After putting on headphones, pressed play, picked up the coils, laughed — was a funny feeling — something . . . not a strong sensation . . . but something, the faintest buzz — an attention side effect?

Felt residual tensions. Am I relaxed enough? Am I doing this properly? Am I supposed to feel something? See something?

What is it like to sync with essential self from a parallel universe? Didn't seem preposterous at all.

When E.J. said to put down the coils, I did, was disappointed it was so short, nothing dramatic had happened.

Sitting there, I had no reason to move, be calm . . . listened to the sound of the water in the fish tank. Watched the changing light levels. People arriving and departing. The nature of my mental chatter had shifted. I was there, observing. Felt no need to get up or do anything else. The sounds of the environment played out a secret place, hidden in the same room. Many minutes went by. Was asked if I was ready to write. Could have sat there longer . . . who knows how long.

Mental chatter happens — don't remember connecting with that still place much recently, yet on both inductions, was left in a very still place. All the things you see and hear are the theater. The play is somewhere else. The play changes.

D. F. — Sacramento, Ca.

I placed my new Beta-Blocker on my kitchen table, hopped in the shower, about 10 minutes after my shower I had a subtle yet distinct sense of an internal good mood while at the same time my beta-mind seemed to exist on the periphery. Truly there was no reason for my mood to shift into a good one. Except there WAS this Beta-Blocker in the room.

Upon retiring, I brought my Beta-Blocker up to my bedroom and placed it on my night table. I then closed my eyes, entered my meditative level and programmed to have a dream. I noticed an ease of entering my meditation.

I must note that it had been a month since I even recalled having a dream. I am happy to say that the beta-blocker feels like . . . Well, I'm very glad to have it. Will keep you updated on this as it unfolds.

I've had my desk top beta-blocker for only 5 days and I can feel the difference. Talk about taking the sweat out of expanding one's consciousness. Wow! Gotta try it to believe it. This is the breakthrough we've ALL been waiting for. Gorebagg's devices are the real deal. No longer are we trapped in this single life-time dimensional experience. E.J.'s transformational work has hit a whole new level. Gotta love it.

S. B. — Manville, N.J.

I have the desk top model. I move it around to wherever I am (within reason) in my place here in lotus land. Last night I put it within two feet of my head while I slept. However, I must say that in the past little while I seem to have become a lot looser in my general demeanor (and inner state). I don't seem to be taking things as "serious" as before. Though I'm remembering to be responsible. One thing I've been adding two and two about is the fun I'm having on the bus lately. (I say lately because it seems the Beta-Blocking effect has been coming through from the station pretty strongly of late, since starting work on the crystal radio.) I'm realizing more now how much the kids on the bus (most of the them are 5 / 6 / 7 yrs old) are also in pretty weird head spaces themselves and I'm beginning to pick up on it more often than I used to . . . even though I was picking it up before, it wasn't happening to me as frequently. Now I'm finding that I have to watch myself and not get too Don Quixotic.

S. D. — Vancouver, B.C., Canada

I have been pleased with the little desk top . . . I sleep with it near (I tend to grind my teeth in horizontal sleep, but this still continues . . . maybe lessened a bit.) When I do any practice I sit near it and have definitely noticed it enhances the stillness of the mind stuff and facilitates the depth of any practice. The help is subtle and can easily be overlooked by the mind looking for something, yet when the "help-effect" is trusted, it is a definite aid in allowing you to get over a threshold and deepen your practice.

M. W. — Athens, Georgia

About 40 minutes ago, my boss and I finished up an appointment between an alcoholic mother (in person) and father (telephonically) in which I facilitated a conversation between them regarding their son. After the appointment was complete and I escorted the mother out the door, both my boss and I heard as loud as any regular conversation from an unrecognizable female voice: "Kathleen, what about your 407?" There was no one there other than she and I and the unit secretary who did not say or hear anything. Both my boss and I heard this voice clear as day. I joked to her, "Wow, there's a parallel universe!" She said, "Yes, I guess so." The numbers of 4 and 7 have been very predominant in my life since I became connected to the school in the late 80s. Don't really know what any of it means other than some form of reminder of the connection.

I do find that since justintv and bb have been such a constant in my life, the effects are multi-faceted, subtle and across environments. I feel the effects in the life outside my home, in each interaction, in the car, etc. "As if" a slow evolution is occurring . . . .reformulating this being called Kathleen. In many ways, I feel like I'm becoming an increasingly simpler version of my core self . . . more pure, clean . . . although those are perhaps not the best descriptive words. I also feel like I have been going through some type of energetic/physical detox . . . the ulcerative colitis has calmed down a lot while at the same time lots of bloating . . . like my body is trying to figure out how to work better.

K. D. — Tucson, Az.

I have been wearing my quarter size amulet for about 2 weeks. Every day at my job it is there around my neck.

At any given time the mental chatter is noticeably subdued, the thinking process is not altered, but it is noticeably lighter, not as dense. If I should decide to pull back altogether from my mental engagement, this device enables me to quickly settle into experiential mode rather effortlessly. It is during this retraction that I can sense the actual field generated in an egg-shaped form encompassing my chest and head. If I am otherwise engaged, the sensation fades into the background.

Last week I came home from work and took off my amulet which I always do. About an hour later I was talking to my wife and told her "I have been searching for this inner self my whole life." Just as I said that, my rather large presence backed up from the machine and entered the space, a few moments later it attenuated back into my body. I did not have the pendant on at this time so I wasn't sure if I could link it to the Beta-Blocker. I had been producing results in my machine before I received even my first desk-top model. And that's the point. We only get what we work for, and everyone knows that. However these devices genuinely facilitate or assist that process.

From the time when I first received my desk-top Beta-Blocker through the present including the quarter size amulet, I have been in a consistently good mood.

I sat in a comfortable sitting position, grabbed the crystal grips, closed my eyes and allowed myself to enter a deep meditative state. Arriving THERE without TRYING to get there.

I turned my intention upon programming for the integration of my scattered selves (or PUPs), wherever they may be. As I was funneling these varied sources and aims into a Central Core, a collating/integrating PUP showed up. I noticed that there was an Integrator of these various energy streams flowing into me. This PUP (form) seemed quite at ease in performing this multi-focused task.

This integrating PUP was sensed internally by me as a multi-focused non-human creature. An insectile humanoid that was alien, yet familiar. It was not a creepy experience, rather, it was benevolent and expansive.

This is my CIA PUP (Collating Insectile Alien) and no it was nothing like the UFO folklore type alien. But non-human it is. And it seems to be able to handle multi-focused tasks rather easily. More than visual I could actually sense the internal functioning of this PUP as an aid in my description of its morphology. I am looking forward to more Beacon Runs.

S. B. — Manville, N.J.

My son just told me that he gave up sugar (again) about 10 days ago and it just occurred to him (as we were talking) that he hasn't been having the usual struggle that he usually did have any time he tried to do it in the past and fingered the amulet he is wearing and said, " Hm, I wonder if it has anything to do with this."

It was one of those moments.

He has been a much changed guy since he started to wear the amulet. I've noticed the change in him. Much calmer; not as stressed; taking things much more in his stride.

I showed him the new amulets and told him I just ordered 3 with a black diode in them at a great price. When he looked at it he said it looks like a baby in a womb.

S. D. — Vancouver, B.C., Canada

I've been wearing my black diode amulet all but one day and night since I received it about a week ago. Before that I had been wearing one of the weaker beta-blocker devices. I noticed right away the increased strength of the black diode amulet. In particular, I found it much easier for my vehicle to be in a dreaming state and the cause seemed to be the amulet. (My intent for the Easter workshop was to unite the dreamed and the Dreamer. I believe this intent is relevant to my experience of the amulet's effects.) I also found myself to be more often aware of myself as an essence (and thus aware of the essential dimension of other people and things) while wearing the amulet. The effect did continue, though, when I left the amulet at home. Thus, the amulet is not like an on/off switch. There seems to be a cumulative effect of enriching the relationship between the dream body and the gross physical body, in line with my intent.

E. P. — Riverside, Ca.

I wear the Black Diode Amulet every day now, and have talked with several folks about it. They seem to understand "meditation aid" and "brainwaves" at least.

Over the course of the last two days, I've received an order for 8 of my wireformed homopolar motor Dragonfly (never been asked about it before), a request for a Tesla CD Turbine (only one other one was a year ago), and a couple of non-toxic auto jobs (I have generally stopped doing mechanical repairs because of allergies). Yesterday, I had the folks at Design News magazine (where I've had 3 Gadget Freak gadgets accepted last year) practically beg me for new gadget ideas. I replied to them with 5 proposals. Since they owed me a check for my last Gadget Freak submission six months ago, I mentioned that in my reply. I then walked to the mailbox . . . and there was the check! I had to quickly e-mail them back and say thanks. So, considering how poor I have been, it was pretty cool to have so many financial offers over a couple of days. But, here's the clincher . . . Design News e-mailed me today, and they like all 5 of my proposals for Gadgets, and more if I can produce them! One of the Gadget Freak projects I "dreamed up" (almost literally) over the last day and submitted to Design News was for something I call the "Spooky Tesla Spirit Radio".

Funny, it was the first gadget they wanted. It's already "slotted in" for the Halloween issue . . . Now I just have to build it! For my "Spooky Tesla Spirit Radio", I plan on using a crystal circuit, but a Tesla-styled design with a spiral-wound Bifilar coil (subject for another e-mail soon). I figure this Gadget Freak will be a back-door way of promoting the "true" Beta-Blocker by exposing thousands of engineer types to the idea that a crystal radio circuit can do many things other than pull in crappy AM. Then, in the build instructions and videos I submit, I will mention my Tesla Spirit Radio idea came from the Beta-Blocker radio, etc, with links to yoyodyne and other sites. And so the fun begins! (or continues . . .)

I was desperate for an air compressor so I can complete a couple of Gadget Freak projects, and knew of one that would be adequate, on sale last week, for half-price at $130.00. It was a cheaper, knock-off version of a better $349 compressor, but all I could barely afford. I got to the store and they said that sale was over. Normally I would be bummed and consider it a missed opportunity. However, somehow I was on a roll with the Beta-Blocker energy, and hung in with the situation, and got this really BIG dude to look it up . . . turns out they were discontinuing the $349 model, and it was on sale for $99 that day . . . but they only had one available! With a chuckle to myself at the perfect synchronicity of the situation, the Big Dude lifts the box (no dolly!) and wades me right through to the front of the long line-up at the till, and even plops the compressor into my car. I was somewhat in shock over the apparent EASE of that situation! Next coincidence . . . I go to Wal-Mart because I need neodymium magnets fast for the 5 Dragonfly motors I'm being commissioned to build recently. I was looking all over for shake-flashlights because they have a good-sized magnet inside that I can extract. Well, lo and behold, there were exactly 5 shake-flashlights left! I bought them all cheaply, smiling to myself all the way to the no-wait (first time!) check-out and beyond. Again, an unusually smooth transaction that makes one feel like "it" really is all happening for the greater good.

Cosmic Coincidence Control is working overtime today! Haha!

R. C. — Sooke, B.C., Canada

At first didn't seem like body relaxed as much as quickly as instructions indicated (even though I have a lot of experience with trance), but I just went with it.

Had a lot of very disparate images, forgotten memories popping up sort of like happens with Spirit Dance. At first was bothered by constant repetition of L315a but then realized it was very specific and deliberate and went with it.

During emerging from session instructions was bothered that there was no completion (or so it seemed) to the instruction to let the body fall like a leaf. Somehow it didn't seem like that one was instructionally reconnected with the body.

At end of cd, sat for quite a while, not sure if integration complete or not. Was much more aware — actually became aware of it when first opened eyes after session for a second and waiting for integration of my body's auric field than usual. And it seemed much "bigger" than usual and more substantial, more "thick". After about 8-10 minutes seemed like a shift had happened and I felt integration had happened and then noticed the auric field was somewhat less noticeable and where it was "thick" before, now it was lighter and more clear.

Now a few minutes later, I still feel "bigger" than had normally in my life.

Comments: Still processing, so not much comments except to say feel excited about the whole process and regaining some more oomph and leverage and aliveness in my life.

M. R. — Santa Monica, Ca.

One of the additional charges I receive from wearing my Beta-Blocker amulet is that it is hugely noticeable, a shock, that I am lugging around a dead piece of flesh. There is not as much, if any, impingement between the HBM and at least a component of the essential self it seems.

My attention can go here and there; there is a grounded feeling. I do not know if it can be called presence, as if someone threw a tramp's wet overcoat over my essence. This is very motivating for a work in progress, like me. Some transformational tools make me aware of states, put me to sleep, make me aware of my sleep.

My Beta-Blocker amulet experience includes a sensation of being a seed inside of something. Through the related practices it is my wish to awaken the machine so that it will function as a transformational apparatus for the essential self.

At the moment I feel the more obvious separation.

"He's alive!" Dr. Frankenstein.

M. S. — San Marcos, Ca.

Mind/Body/Spirit state upon start was one of barely contained chaos. Had been working with a lot of charged energy successfully throughout weekend, but by point of session, a breakdown had begun and was on verge of completely blowing the energy build-up. The space was a bit noisy, but once I turned up the volume and focused on the induction, this noise went to the background. Despite atmosphere, both internal and external, which produced great challenge for me, I was able to achieve somewhat of a relaxed state, enough to feel momentarily out of body at the corresponding point in the induction in which the body is dropped. All sound and movement seemed as if from a distance below. It seems I see energy in color — last induction I saw the infusion of red energy, this time it was purple, a beautiful brilliant purple. This triggered the memory of my morning's work with the *Angel's Healing Journey*, where the radiation was orange and the entity, archangel Suriel. With this came orange into my field of vision, but not as brilliant as the purple. I noted that while I still maintained a needed posture of self-containment, there was more ease to this, as though there was added support and strength, and the mental considerations had calmed down considerably. By the time I left there was the experience of being closer to sufficient unto myself, I needed nothing from no one, there was a feeling of impartiality and definitely more confidence. I longed for the pleasure of just my own company, and am enjoying that Presently. Thank you very much for the opportunity. I hope to get to the place of owning my own Beacon as I would love to explore the possibilities that might present themselves in a more controlled environment.

K. R. — Penn Valley, Ca.

After the Beacon Session I felt completely relaxed, refreshed and renewed, and happy. I felt that my whole consciousness had been expanded and there is a new dimension available in me now. My energy is excellent, 4 hours after the session — I'm feeling more vibrant and stronger than before the session. My outlook is more positive and flexible and I look forward to the rest of the day. Optimistic and fluid would describe my new mood and more confident than usual.

It isn't every day you can go into a shop on a break from work or shopping for 15 — 20 minutes and sit quietly tuning in to an astral travel adventure that changes your emotional state, your mental state and your cellular composition and leaves you feeling like you've just been dancing among the stars.

Well, that's what happened to me this morning when I stopped at Brane-Power for my Second Induction training. I still feel the transcendent waves of bliss that are moving through me. I can't explain it — it seems to be the natural outcome of a very simple process of holding on to two tourmaline crystal grips connected to the Beacon Strapper and listening to a taped CD of a voice suggesting deep relaxation (ahh, you mean I can just totally let go and relax?). The voice describes being in a space in a parallel universe where I can receive something important to my essential self. This is powerful stuff!

I think it's rubbing off on people around me — everyone seems terribly happy today. I can travel to a parallel world in minutes instead of waiting for enlightenment to penetrate through my thick karmic skin. Hope is just around the corner! Thank you.

L. N. — Penn Valley, Ca.

The effect of the Beta-Blocker amulet was very noticeable in the beginning:

— The first night I slept with it and I had a lucid dreaming.

— Some coincidences happened, or I noticed them better, one of them was: "I had to address a question to the administration bureau of my school, but somehow I didn't want to start the conversation. The next day I entered in the administration bureau and they were arguing right on the subject I wanted to ask about."

— The first time I took off the Beta-Blocker amulet after wearing it one day was clearly outlined what it was blocking: the inner speech, that I know quite well, enhancing with imaginary stories the deeds of that guy "Jorge".

J. P. — Portugal

When I first got the desk top Beta-Blocker it was just a very incredible powerful experience. I do notice you must focus on the desk top Beta-Blocker to somehow activate it or help it Beta-Block.

Sometimes I feel like I went through a "mini dark Night of the Soul" with the Black Diode amulet, mini in that it seems to be over and I seem to be through the other end. During this short period I almost wished I had never started to use the amulet because of what I am going through, but I know I'll have to go through it, once again if I don't go through it this time, so the way out is to go through it". During this period the amulet felt like a burden around my neck.

Today I got the crystals and the cables for the Beacon, and in a curious sort of way it seemed to "Complete" the Black Diode. I look forward to further experimentation with the Beacon and Black Diode.

I find it much easier to get into Delta, where my body, machine is asleep, completely relaxed and I am awake.

Dreams — I have this one dream that I have had lots of times, but can't quite remember it until the next time when I remember that was the same house that I dreamt about last time.

It was also very helpful E.J. short video on the Chronic and how we may think that the amulet is not having any effect; it seemed to suddenly make its effect much stronger.

I have noticed that I worry a lot, about things that haven't even happened and that may never happen, in L315A anyway.

Ah Prosperity . . . well having had a tough year, business seems to be picking up, and I am actually doing better than I thought . . . but I am a bit paranoid.

J. B. — Vancouver, B.C., Canada

Just a quick note about the effects of the black diode and the beacon. The office at the Arts Alliance has become an ongoing beehive of activity with little work clusters of artists working on several of our major projects and a constant stream of drop-ins. On Sunday we had the unveiling of our public art project in central park 48 artists with around 100+ community people, laughing, enjoying the art and being with each other. Even the "die-hard" and the disenchanted have come back to the fold. I feel like I'm working at hyper-speed, but in a smooth, rippling, productive way. Here is a comment from one of the artists who has never been a big fan.

"Tamara, I feel like today was a whirlwind and judging from what was going on in your office I think you must be wearing red shoes and be in a tornado somewhere. I really appreciate you."

T. M. — Rowland, Pa.

Lorraine, Val and Kelly came at around 10:30 a.m. this morning and after each had sat with the Beacon for at least a half hour they all did BSI L315a. Kelly went first and reported that she could feel a shift and a grounding and that it was definitely very extraordinary. She also commented on the quality of E.J.'s voice and compared it to other Alpha cds she has been listening to and much preferred E.J.'s voice.

Val was next and she burst out into a radiant smile towards the end and had a very joyful time of it. Much happy laughter ensued. She also reported seeing many colors.

Lorraine then did the strapping and it was very successful. She talked about it being very grounding. However, we did not discuss it much after Lorraine's induction.

We then went for lunch across the road and wrapped up the session.

They all want to do more inductions and both Val and Kelly were intrigued by the whole thing but thought it was great.

Val is from Kelowna (the third largest city in B.C. — in the Okanogan) and is just down for the weekend. She works in a care center. Kelowna is known as a place where old people go to retire. It is on the Okanogan lake in the Okanogan valley.

It was quite an interesting morning. While all this was going on the mailman delivered the rest of the Induction cds.

S. D. — Vancouver, B.C. Canada

I had a great Beta-Blocker experience at the chiropractor's today. I wore the Black Diode Amulet and showed it to the doctor. I told him about the good fortune I had within days of wearing it, and that I created a great new job making animated wire-formed dragonflies because of it. We ended up talking most of the visit. He was excited and suddenly offered to exchange the next chiropractor visit, "or two", for one of my dragonfly pieces, sight unseen!!

The transaction was effortless, and I left the office feeling far better than normal from the chiro adjustment.

I will introduce him to a Beta-Blocker Desktop next visit . . . it's "Absolutely" perfect for a chiropractor's office!

R. C. — Sooke, B.C., Canada

Participants: Koyote, Eric, Jenny P, Liebre —

Voyaged through various spaces as though we were in a helicopter — 6 directions. Color changes around temple door indicating blues, oranges, purples. Buzzing noises and high-pitched streams. Light changes in the atmosphere and electrical currents and fog present in moments, mostly in mass formation. The room moved and shook as though there was an earthquake. The space also moved 180 degrees on one occasion. Perceived Koyote took us to alternate universes where other creatures/beings were present. Saw the nagual scratch through the temple wall (nail marks). Also viewed 2 huge horse flies the size of large pigs which were collared and leashed. Another presence was steering them to my left. Experienced the sense that some of Koyote's, JP's, and my PUPS were retrieved successfully. When Koyote piloted, I smelled cedar/douglass fir/cedar tips. It was as if we were traveling just above the tree line where nothing can grow but pines. We seemed to voyage pretty far out there. The space seemed as though it has vaulted ceilings and the eyes averted upward and attention diffused as it so happens in Saint Peter's Basilica (in Rome). A sense of holiness as though we were in a temple or grand church.

The experience I had with retrieving my PUPs was that of a water snake that slithers slowly, methodically, and gracefully through the water. I felt as though a leech had sucked things into my belly. Recovered memory of feeling that I lost something that left my belly button one month ago when I was dreaming. My sense is that this was my PUP and that day I retrieved her.

R. F. — Riverside, Ca.

Lorraine bought one of the Twin Turbos from me to give to Bo (A Chinese student at the t'ai chi club — he is a long time student of Master Chung's and a very nice fellow but not very "prosperous" if you will, and a few weeks ago was hacking and coughing and telling us he had TB.) Lorraine gave the Twin Turbo to him as a present. She was telling me today that he told her that he didn't wear it to bed till the third night, but when he did he got a full night's sleep for the first time in years. He also said that he is not bothered by all the bullshit around him at work anymore and he's feeling much calmer generally.

S. D. — Vancouver, B.C., Canada

Just wanted to give a little feedback concerning my donning of the Twin Turbo amulet. I began wearing it on the same day of my first induction, so my recollection of any effects were initially affected by the Beacon session at the Brane Power Center. Following the session with the Beacon I began wearing the amulet and noticed feeling a bit high or "out there". This was accompanied by a lot of sixth/seventh chakra activity and energy movement near and around my head. That first night it was difficult for me to sleep and I wasn't sure if it was due to my excitement and upped energy from the Beacon, the amulet, or just me and my own mind. But the following day I felt calmer and less excitable while at work and this continued throughout the week. Overall, sleep in the night has not been as difficult as it has been for me in the recent past. I have noticed less of an inclination to flare up during stressful moments, more open and friendly with co-workers and clients, and a sense of calmness that pervades experience of day to day life. I also completed my second Beacon session and though initially less intense, I did notice a feeling of being more relaxed while the induction directions occurred.

D. B. — Grass Valley, Ca.

Thank you for this marvelous product (bought a twin phase amulet). I do wear it all the time and found out that I get a restful night's sleep and during the day I feel very calm and relaxed. It seems to suppress anger, aggression and ego. I think kinder and gentler thoughts. Thanks again.

B. L. — Vancouver, B.C., Canada

Mr. Macho Beta-Blocker presented the palms of my hands with force from both the crystals and the coil. I could feel it coming from both. Then there was a jolt in the palms of both my hands near the base, up my arms, a flood of nervous energy across my chest, not just simple tingling. Then my solar plexus expanded with great sensitive energy and on to the pelvic furnace, wham, down my legs all the way to the skin on the ends of my toes. Then I noticed my head at the base of my skull was full of energy as well as the back of my head and then I felt it move around my temples, through the center of my head, forehead and into the crown. It was then that I realized that I was CHARGED, very nervous and had to drive all the way back to Sacramento. I did not believe I would experience such an event. Not a light and sound machine for sure. I am stunned. I should be able to do great things with this technology in view of the waking visions and very busy astral at night. I can't wait until I have an opportunity to begin my sessions. I was zinging from crown to the skin on the ends of my toes. What a great day today! Thank you.

S. E. — Sacramento, Ca.

Spirit, my dog, did not like it when I took her collar off; she looked horrified and watched me carefully as I took it to measure it. She followed me until I put it back on her, then she practically jumped into it!

The other day I heard her making this odd sucking noise and went over to see what she had gotten into now. Well it was the fiber art pin that is on her collar. It had come off during the night and she was holding it in her paws and sucking on it like a pacifier, very loudly too! LOL! I found the pin part and took the fiber part to put it back on and she wasn't sure if she should let me have it. She watched very closely as I put them together and then when she realized I was putting it on her collar, she finally lifted her chin to let me. Since she got her own little fiber art (to go with her L1 AkaDua) she leaves my medicine/energy pouch alone! It will be interesting to see how she takes to the Beta-Blocker. Next she'll be wanting her own Beacon! LOL! Well, she is aptly named for voyaging: Sojourner Spirit.

S. C. — Redlands, Ca.

My amulet is very interesting and the only thing I wanted to do all weekend was sleep. My mind has become very relaxed and my intuition is much more accessible.

D. S. — Baltimore, Ma.

I sat with the Beta-Blocker for 3 minutes and after about a minute I realized how much tension was in the back of my neck. I started to feel more relaxed right at about 3 minutes. I felt more centered (as in meditation) and relieved of normal everyday pressure and tension. If the Beta-Blocker can get results from a nervous wreck type-A New York personality like me, I guess that's saying something.

A fly landed on my nose while I held the coil-wrapped crystals in my hands. I didn't even have the urge to swat him. Anyway, if I did, I would've broken my nose with the crystal!

J. A. — Penn Valley, Ca.

I've been suffering from insomnia where I wake up after 4 hours of sleep and can't get back to sleep the rest of the night. The first night with the BB in the room next to my bed, I woke up as usual but was able to get back to sleep for another 3 hours. Great!

Mick P. — Baltimore, Maryland

I have been using invocational techniques and the Telling to make the voyaging available to others, and now I believe these apparati will provide a great navigational tool that might allow me to bring the Telling to the next level by applying the shamanic functionality of my assigned voyaging pod (i.e., biological machine) as it has been transformed for shamanic work as the required membrane and directional navigational computer with the use (of course) of your navigational tools. From a shamanic point of view, we've been waiting for these tools for nearly 26,000 years. From the Toltec point of view, we have not seen these possibilities for a thousand years. From the point of view of a Salvadoran voyager, every great inter-dimensional voyage of a thousand light years begins with a 3 hour delay. So, you are right on time, E.J.!

K. — Riverside, Ca.

The inner chatter is not gone, but it is easier to fade it into the background.

The first couple of times I brought the table top model to band rehearsals were the best rehearsals we've ever had.

Everyone seemed more relaxed and playful than usual, especially the ones who usually are a bit edgy. There was more jamming than ever before — before, in-between and after the songs — on the set list.

Not many band members took seriously, or even listened to, my simplified explanation of this "chill-out radio" as I called it, but one who did and another one who didn't both told me that it seemed to work. Our youngest guitar player requested that I bring it for every gig and rehearsal.

The second gig we did with the BB (as always on top of a large speaker) was i.m.o. our best gig so far. Several ladies in the audience were moved to tears that night. We were contacted about playing at the largest TV charity event in Norway as the only "unknown act" after this gig. (We'll be playing it in October.)

One thing I've noticed ever since I got the table top BB is that it frequently makes the world seem vast and spacious, less cramped and claustrophobic. The first couple of times I had it in my car while driving to band rehearsals, I felt like I was cruising through space.

The BB seems to make it much easier to go into and maintain a sort of "waking-dreaminess" and a feeling of being something lighter than the physical body.

Judging from the behavior of people around me, I seem to inhabit a friendlier world than I did just eight days ago. As when I got the magic charms, I seem to notice a change in how people behave towards me. The amulet seems to take it up a couple of notches, so to speak.

H. J. — Oslo, Norway

Please let E.J. know that the prosperity amulet is working in profound ways. I have been working night and day to come up with sufficient funds to prevent losing my home of 28 years in a tax sale if the outstanding property and school taxes were not paid. Paying them in addition to the mortgage I pay has been extremely difficult since my husband died last year. I wasn't sure I was going to come up with sufficient funds doing the hospice and eldercare work I now do. I called today to find out how short I was going to be; it turns out that the school tax had somehow been paid, I have no record of that. Now I can fully pay the tax, and my home has been spared from tax sale. Thanks.

B. — Pa.

Thank you for introducing me to the Beacon today. It was an interesting experienc. This is approximately the set of thoughts / associations I had with it — they proceeded forward and backward:

My attention was drawn to your mandolin and to the eight strings in four pairs. They looked as if they would vibrate together, in pairs. The Chinese traditionally fear four, representing death, and are attracted to eight, representing good fortune. The numeral 8 is the same as the symbol for infinity. When I halved four, I got two: duality. When I halved two, I got one: unity.

When I doubled eight to sixteen, I doubled infinity and got another universe. And doubled again and got two more, and doubled again: with every beat of my heart, it doubled and when it became sufficiently immense, it became one again.

It was the same, oneness. It kept transitioning back and forth between the two: duality again. I'm still resonating with it. That's all I can tell you. Thanks again.

M. M. — Vancouver, B.C., Canada

AAAAAAHHHHHH!!!!!!!!! My prosperity amulet arrived this

afternoon.

I put it on and was writing a little thank you email to you guys when I got a call telling me that my father's estate has a check for me in the amount of $150,000! I had been wearing the amulet for maybe eight minutes!!! We, like so many people these days, have been really struggling to make ends meet. We were even wondering if we would have to sell our home. I borrowed money to buy the Beacon.

Parenthetically, I was having a very anxious night last night, so I was reading the ABD for quite a while to become present and dispel the illusions of fear.

When I did get to sleep, my dreams were of E.J. in an old New York apartment. We were sitting at a kitchen table and he was teaching me to draw and telling me these hilarious stories, none of which I can remember now.

Please give E.J. my love and gratitude. What a blessing to find such a guide.

M.L. — Riverside, Ca.

I am having a lot of fun with my Black Diode Amulet, and my connection with the Divine is getting clearer and clearer each and every day. I now have crystal clear connection even when I am not wearing it. I have also noticed that after I received the Black Diode Amulet, I am able to smile and be happy even in the midst of conflict and turmoil. I smile a lot more than I used to, almost all the time. And I have also noticed that conflicting situations usually resolve almost automatically by itself, without me having to do anything; this has never happened before, but now it happens all the time, which is absolutely great!

I believe that through using it on a daily basis since I received it, I have come into contact with several of my parallel selves . . . .

I have gotten phenomenal results. Much better than I hoped for actually, so it's going to be really exciting to see what results I will get with the Prosperity Amulet and the Smart Pill Amulet too.

B. H. — Oslo, Norway

Sat with the Beacon for about 15 minutes. I did feel a pleasant serge of energy that vibrated through my palms. I think I was calm when I came in and I was calmer and felt more balanced when I left.

Everything (time) seemed to be slower to a point where time didn't seem important. Respiratory rate, pulse and thoughts all slowed to a calm being. It felt good to be internally quiet.

In the present moment . . . clarity of mind and a realization of that nothingness between thoughts. There is a peaceful feeling toward a hint of joy.

This is a gift that I am so grateful to have used to remind me of my human/spiritual beingness. Namaste.

Put in CD, listened to Instruction, following instructions and counting, breathing, I started to shift. By the time I got to the count of 5, I noticed the speaking of the words becoming difficult. I could hear them, think them, but the speaking felt disconnected. I noticed that there was one me observing the "me" trying to speak and a 3rd me, the physical me, that felt heavy, sluggish, and immovable, like I was trying to control and manipulate someone else's body as I was outside of the body — my body — that despite feeling disconnected I still recognized.

It was a very intense experience, both in observing/participating and in the physical effects afterwards. While sitting I felt my body move several times, sliding/falling sideways, and a couple of times I felt myself black out. While I encountered struggling to gracefully shift, I also noticed that my consciousness (thoughts) were awake, lucid and clear and a change in intensity of/in this process, different from when my eyes are open and I'm voyaging.

Listened to the next CD, the transition was smooth, relaxing and invigorating. Following instructions, I felt myself moving through the space/states. I was able to recognize the sensations and noticed that this time it was smoother, a little easier, possibly even faster. This time I felt a slight, short time of resistance to the instruction, pulling on my attention and settling my body. I was able to move past and beyond the resistance and settle into shifting through the space. When I opened my eyes, I did not experience this time the flash/blink of slight disorientation and/in different space.

I felt more refreshed after #2 than the 1st time. Each state seems to affect me differently and voyaging gracefully will take much practice. I really enjoyed it!

S.C. — Redlands, Ca.

I was lucky and got to go to the Brane-Power Center to experience my second induction on the Beacon, strapping to universe X-967E, and to my corresponding PUP (parallel universe persona), and then integrating it. I found it easy to fall right into Delta (just for the briefest moment it seemed, but when I was told to awaken, I realized I'd just gone way down into sleep, but didn't notice doing so. Also, I was aware of some resistance to the situation of the body going numb, although it allowed a touch of it, so to speak. Funny I never thought about the back of my neck, just about ever.

I sure like this universe, or maybe I'm the pup, saying these universes are delightful — having been in the other one? I'll remember its name is X-967E. Maybe I'm just integrating the PUP from there to here — I get them all mixed up, but the center was a great atmosphere for the training.

Tell E.J. thank you!

J.H. — Grass Valley, Ca.

My intention was to integrate this PUP into home universe.

Relaxed the machine. Felt the crystal grips warm at my touch. Quieted all in the machine.

When focusing on securing the PUP at X9-67E and identifying which PUP I believed it was, my heartbeat sped up and my hands began to sweat.

Realized that this PUP has been lost at X9-67E for some time (injured childhood "I") and yearned for integration, attention, conclusion. After integration, concluded that this PUP was now found, secured, and integrated into my universal persona. Emotional swell!

M. M. — Grass Valley, Ca.

Experiment using machine as a part of work on self.

A quiet voyage to gather a part of me that was very near. Integration felt like a gentle inhalation of fragments of myself. The voice of the guide was easy to follow; the process effortless.

A focused deliberate exercise to accomplish something that I seem to have been doing. Being conscious of and guided in the process. Made it seem easy and very natural.

Those machines can greatly aid elegant voyaging.

I'm wearing the "Oso-Double Black Diode Amulet". Wow! I was immediately aware of the levels of my energy body. 1. Solar Heart. 2. Body Core, 3. 2 ft. Field Around My Body. Movements and currents were very distinct and clear. Found the guides easy to follow.

This technology seems to create the conditions to complete energetic work that would otherwise take place over very long periods of time (if at all). The guided instructions were like the gentle voice of my guardian angel, helping me to complete necessary steps in the unfolding of my spiritual path.

The Beacon clears obstacles, E.J. offers guidance. It's still my work. E.J. Gold is a high level spiritual engineer, developing and refining a technology in the service of All Beings Everywhere.

O. — Riverside, Ca.

Began to voyage thru different earth environments via olfactory sensations — smelled the spaces of ocean, swamp, a metallic world, a world where I smelled the hard work that comes from human sweat. There was a sense that I was also on the People Mover or Space Mtn. Disneyland . . . the slow smooth turns and stops that happen as well as the accelerations.

I went thru some spaces I visited as a child when passing thru the bardos. I saw thru shadows closeby, as if a person was walking in front of a movie screen. The shifting of the body happened. Felt like I spent a lot of "time" in the void. Felt at home and familiar.

Strangely, I passed thru a space where a movie star had been. He had a pig nose. Also, in one space I myself with glass-like eyes & an ice pick slowly penetrating thru my left temple to my right temple. I also saw E.J.'s BardoTown — his city lights and curvy road into the void.

I felt like I was in communication with another intelligence. Listening to E.J. music created changes in me whereby it was produced. I felt like a sonar communication happening. I felt I was receiving a signal and sending a signal. The music that sounded space-like made me feel I was a home. I remember stars moving in their own orbit and then dancing together.

I believe music creates a different type of vehicle from which new worlds can be explored. It seems that music holds codes to me — representations of real worlds & spaces each with their own laws and physical properties. I would like to explore.

K.H. — Riverside, Ca.

My intention was to retrieve a part of myself from another parallel universe. Right at the beginning the first thing I saw was a very wrinkled old man with wire-rimmed glasses. I saw his left profile from the neck up and I could see that he was smiling. He felt like a scientist or an inventor of some sort. Then there was space with varying degrees of light. At the very end, I saw a little boy from the back in shorts and t-shirt (orange shirt, beige shorts). He was reaching into a big box, on his tip-toes, to pull something out. There was an older presence watching him. A couple of times, 2 or 3, I'd seen the front view of the alien face — a round one.

I think that the old man and little boy were the same being. And from the end of his life backwards to the beginning is what I was getting. And I hoped that the little boy was the one that conversed with wild animals when he was little.

This is only the beginning, and it's the end too.

My intention was to experience the traveling to other worlds.

As before, I experienced a buoyancy when getting into the deep Delta. The pain, physical, is no longer felt. The murmuring in the room seemed to come in waves and coincide with what I was experiencing — the rhythm of it. Just as the voyage ended, there was a burst of laughter. A great welcome back!

I wonder if this could be used as a way to decrease the physical pain folks may be feeling in their bodies. In a very real way this experience felt a lot like being in the tank, the Samadhi Tank. Even as I write this and supposedly am back, I still feel not quite back — like if I was in a bubble aware of what's going on around me, but not part of it — not quite. The smells in the place are intense.

P. — Riverside, Ca.

Level 1, #1 — I went into sleep state, felt quiet, deep breathing, was able to block out external noise, deep relaxation.

Level 1, #2 — Feeling a slight bit heavier & relaxed. More quiet inside.

Level 1, #3 — Felt quiet and uplifted-energized. I don't quite understand it but appreciate the mystery behind it. Thanks!

R. U. — Nevada City, Ca.

Got very quickly and deeply relaxed. Saw vague shapes and had the sensation of traveling through them. Similar to experiences in the floatation tank. Calm and clear afterwards.

O.F. — Grass Valley, Ca.

I fell into a deep, deep sleep. I remember being in that state and knowing I was there, using my body as a gage-limp, deeply relaxed. I continued to listen to the instruction, but when it came to meeting my other self, I seem to have lost beta-consciousness blackout.

I want to try it again. The instructions said I would be conscious of my meeting, but I don't think I was. I was conscious of entering into deep Delta.

C.F. — Sacramento, Ca.

When the CD said, “You will now enter the parallel universe”, I felt as if I was invisible. I could see everybody but no one could see me. Then I felt like I was traveling backward and I was like air; I had no shape or form.

I started feeling very sleepy and my hands were pulsating as I was holding the crystals in my hands. There were times when I did go to sleep and when I woke up, I felt like I was in a different world.

A.S. — Riverside, Ca.

L315A — It's so quiet here. How can it possibly be so quiet? Grounded, heavy, yet soft rooted, deep, deep calm. First integration into the brain infused, but not yet grounded. Is it done? Move into the heart. Second integration blends into the root chakra. Ah, now the centering occurs. Awake. So quiet inside the strength of self intensified. Want to stay here . . . feels good. Feels right. But how now to infuse with the environment outside self? Oh, I am so fully infused I just need to take the breath, lift the eyes . . . make contact.

A.H. — Grass Valley, Ca.

The grips were immediately warm to the touch and heavy with a warm liquid sensation. The body felt instantly enveloped in a warmth. There was a certain amount of anxiety about what would happen, and if anything would happen at all. A lot of external activity occurred while the induction was happening. Loud talking and movement all around which made concentration of utmost importance. There was a sense of a largeness hovering or coming into my space/body and a moment of dismay and fear, but then a feeling of joy like what I have known as a young child. Then there was the sense that this was all taking place in something like Grand Central Station (a lot of loud talking and quick movement in the induction space). When I thought "this is an overlap with L315a space", I was able to proceed with the induction. It seemed that a wonderful magical thing was happening right under everyone's nose and they did not even know. Related to both the induction space and L315a — a mischievous, playful sense of what was happening. An electrical feeling but not tingling. Quiet, calm, strong. Happy to be with her/him/it. Wishing not to leave to know more to have it last longer. To take it in more completely. Not really knowing but sensing such an amazing kinship with something/one that I cannot see or feel in the homeworld way. This could be an addictive thing. When it was over I did not want to move or speak. When I did rise from the chair, I was as if not connected to the body and walked a little as if listing to one side or not quite touching the ground. I've typed this whole report with my eyes closed as when I open them, the keyboard, hand, etc. seem alien or as though not connected to me. They seem a long way off and there is a feeling of amazement that they can function on their own.

B.H. — Penn Valley, Ca.

Honestly, while I was open to the experience, whichever it may be, I did not really expect anything much to happen except some relaxation.

I did a couple of somewhat deep breaths while holding the probes/crystals. Then placed the little piece of paper with the coordinates on my desk, another breath or 2 and murmuring of the coordinates.

Then a moment of waiting . . . a couple of minutes, maybe occasional shivering, twitching unexpectedly, about maybe 5-7 meters in front of me, about 1 meter high, a female creature appeared of about 5 feet tall, old . . . whitish gray, almost raggy dress, silvery-gray hair, about shoulder length . . . the word banshee comes to mind, not really hag, not a witch . . . well, maybe just this woman.

Not sure why I started crying, just a few tears, some shivering and trembling, not sure if someone sitting there would have noticed.

I could see her walk, mutter, make sounds; her environment was that of a somewhat wild forest. I looked at her and how she moved . . . the sounds she would be able to make were there, although she did not actually make them — they are intense sounds, some shrill . . . there may be, for humans, a crazy quality to them as well.

Then suddenly I am looking down at my feet and onto the forest floor . . . look over my shoulder and see the same clothes she is wearing, not the blue knit sweater I sat down with. I am her and things seem real, am interested to hear or sense the forest floor. I am wondering when the endpoint of this would be . . . I tell E.J. about my experience and he tells me to get acquainted with her and then send it and put the probes away. So I sit there some more . . . she and I are not separate, I sense her as me, she is slightly smaller.

I know that she rarely communicated with humans . . . but "talks" to the dead, or disembodied spirits. She knows the way of the forest and the creatures very intimately, mushrooms included, potions included. Sometime she may be called upon by what seems to be humans, the sounds she makes are unusual, and I think could be scary and might be regarded as coming from someone crazy if heard. I actually feel tempted to physically make them and I know I could. It feels I pulled her in, not sure what that means though. I end the session after a few more minutes as it feels a certain point of integration has happened that seems enough for the moment.

C.W. — Penn Valley, Ca.

The body became very relaxed.

Every time an image started to evolve it was resisted, did not want to use my head/visual for experience, resisted about 4 or 5, then fell into a very large sensation or the sensation of being very large, maybe about 30 times this size. The only thing I could see was a huge head, mushroom-shaped, sort of tapered at the top. Two short legs with feet. Like a two-pronged mushroom plug, but soft and mushy, just floating in space, with the "impression" of eyebrows and eyes, no color; so let's say whitish beige, on the dull side. I could hear a heartbeat at some point.

I felt a huge hole in my chest and then I felt "pain" in that area — a different kind of pain. A feeling that you did not want to continue — more toward emotional pain, but not the tearful kind . . . maybe too hard to explain, but I could not take it for very long.

I could not put down the grips when the voice said to. I sat with them in secret, just sitting and that is when all of the above happened.

I was blasted when I opened my eyes and everything seemed to be spread evenly, so it is difficult to settle down on what to do first. I did not want to write the report and did a number of things and then I was here writing the report.

Y. — Penn Valley, Ca.

Without the focus of the induction CD, I noticed more mind wanderings, flighty thoughts going around to nowhere. Near the end of the 10-minute session, I noticed sensation in my hands (similar to the sensations that have been present in other sessions). At the conclusion of the session, when I opened my eyes, it was startling/surprising to find myself in our Beacon room/center. I am taking all the "chronic" information within, pondering/holding this perspective of the chronic and its relations to the Beta-Blocking devices.

With this format for logs, I find myself writing for an audience, rather than for myself. Field Notes v. Field Reports.

My intention is to keep a regular practice/observe/be open to possibilities. The establishment of a consistent practice includes:

a) relaxing of the facial mask

b) placing attention at the base of the neck, where it connects to the shoulders

c) radiating the sensations outward, from the base of the neck, towards the remainder of the "human body"

d) observing and noting and "phenomena"

As an indicator, I tend to look for the sensations in the hands as a sign that I am "in tune" with the beacon transmission. This sensation was the first "difference" I noticed when I held the grips for the very first time. This sensation has been less apparent during the subsequent uses.

Noted in this experiment:

Mental wanderings: Wednesday's Diablo safari, various struggles with money. This led to other wanderings about a defense of hibernation — closing in & building a wall between the experiences of this life orb and the pain and fear of "things not going very well". Wanderings & Searchings for an experience that lends to the realization of being connected with the other PUPs. Attempting to open the heart, passed the hibernating impulse to withdrawal to a safer distance, not to mind the pain. Weary of self-pity, and wishing for the innocent enthusiasm. Is my "luck"/magic running out?

Finding the 15 or so minutes a "nice" exercise in attention, similar to the series of four meditation workshops held a few years ago.

Waiting for something? Looking in the wrong direction? Closed to possibilities? Nevertheless, the practice will continue.

F. S. — Chicago, Ill.

I hold the crystals and immediately notice the tension points of my body while feeling the pulse in my hands.

Dropping the facial mask, tension in general. Easy enough (maybe because it's the first thing I do after waking up and I'm still half asleep).

I close my eyes and see circles and waves. I try not to choose intentionally anything, just let whatever come.

I see "clearly" a place I know in Spain, by the sea. First I want to go somewhere else, but something tells me that it is the right place. I let myself go there.

It is not easy to pull my persona from there. On the contrary, it seems that I want to go there fully.

The time to pull my persona into here, this world, is too short for me. I take a bit longer after the cd is over.

When it seems to be enough, I let go of the crystals.

I lie down on the floor, facing up, intuitively massaging my belly. Somehow it's helping me integrate whatever has happened.

When I leave the room I feel totally refreshed, full of energy and willingness to do many things. I'm aware of my body in a special way.

H.S. — Chicago, Ill.

Picked up crystals wrapped in copper.

Instruction to relax all muscles in body, by first locating spot of numbness on back of neck, and then allowing it to spread through body mass, such that the body could go down into deep Delta sleep state.

"Seek you" was repeated several times, or maybe it was "C Q" to call the persona, and allow the beacon to guide it to L315a.

(Note: did not go completely numb, still noted the Beta-Brain noise as faint thoughts about shopping for medicinal oils today, due to ongoing body/health preoccupations — in other worlds?)

I noted at least 3 personae alerted. I tend to visualize them as my actual friends . . . not mirror people, or copies of me . . . feeling uncertain. Definitely went into very pleasant alpha . . . this is a very nice and gentle state, hands warm up and kind of pulse.

J.H. — Penn Valley, Ca.

Easier to travel with the induction instructions on my second voyage — moved into the space quicker and with less resistance. A very short while after starting the induction, the grippers became very hot in my hand. As I traveled into the space, I was aware that the space was an energy vortex moving in a counter-clockwise direction. The persona of L315a was sheer energy itself. I opened myself up to the flow and felt myself join with it, become a part of it. Again at the count of 3, I awoke refreshed and not tired at all and eager to write down my experiences.

I want to come back again and again to the space to fully integrate with it. I feel it is a life source which will nurture and energetically sustain my essential self. This space is fuel for my Being.

J.M. — Penn Valley, Ca.

My body followed the instructions to relax without reaching the sleeping state, while the mind was very alert. When the instruction was given to find an alternate persona, there was a sound of an engine revving in a race track and I found myself climbing into a race car as the driver in what seemed to be a racetrack during a race. I spent my time adjusting my goggles and climbing into the car and then I came out of the state. I feel very alert, calm and steady. The race car driver persona seems a new idea — neither attracted nor repelled.

L.P. — Grass Valley, Ca.

At first the grips felt icy cold although they warmed up quite fast. I was surprised that the time of the induction seemed quite short. Even though it seemed short, I still felt that I was able to follow the instructions and complete the assignment. I seemed to have various body sensations such as tingling in my legs, temperature change in my legs only, vibrations throughout my body. At this time, I am feeling a bit lightheaded and a little shaky.

I believe that I did reach a deep Delta state in the induction. I am not sure a persona actually strapped on, but I am willing to take the information in and let it be for awhile.

MA. W. — Penn Valley, Ca.

My intention was to experience what my path is revealing to me as it is now being more fully realized as of late. To connect more strongly to myself so that I can carry out my purpose far more efficiently.

I was and was not surprised at how quickly I could drop much of my face mask, and how gradually it dropped further and further off my face. I had a bit of worry as is normal with me as to whether I am experiencing correctly or doing it right — that I released without strain, and I felt my body relax fully.

My consciousness as I understood it still felt primarily in my head area behind my face, and trailed, thinning out at the extensions of my limbs. I knew not what was happening or what I felt; a couple of times I thought to move a finger or my mouth, worrying about drool or maybe realization that my body was still connected to me, but realized quickly that to do so would take much effort and that this effort could cause me to abort the mission.

I felt a need to sneeze brewing in my nose, and thought "interesting" when my body did not only not respond, but also drifted further (or felt further), the skin a shell. I felt myself connected to another self that I have normally felt as possibly an imagined archetypal, hidden, unrealized or forgotten self who was and yet was not me. I felt it as if behind and in front of me simultaneously.

When I fully stopped my thinking (what?!) or (interesting . . .) or (when . . . ?) I hear the voice again; my body is about to awake and my eyelid flutters. At the snap my eyes open. Just before, the words "energetic" and such registered as meaning differently than what I actually felt. Then without realizing it with words, I grasped that my received notion of energetic pertained to that which I could not fully connect with something I had been told. And then when I opened my eyes I felt a serious excitement, which reminded me of my earliest childhood. An honest smile swept over my face, honest because it was fully for myself and for no one else

When I waited for full realization, I felt that a spot on my lower back was tunneling in a kind of white light. I was growing tired of holding the crystals, but could not drop them. I would focus on that center on my back and could see how much of the spiraling tunnel was left until it was all in my body. A similar silent movement occurred when I was able to let go of the crystal as when I came in connection with my parallel self. When my face had dropped, I felt my face look like my father's.

X. F. — Riverside, Ca.

During this training I heard differently. When the numbers were given for the parallel world, I heard them for the first time as coordinates. And I would feel my attention, focus, slide down an axis when the first number was given and then when the second number was given, I felt like a jump occurred and I was in a completely different location and I had followed the numbers to get there. Every time the numbers were given, I felt a slide and a jump and a sense of arrival.

At the conclusion I felt happy and relaxed and full of energy.

R. G. — Grass Valley, Ca.

First experience with the Beacon plugged into the wall. I was in the dining room and the lights were off. Few people were in the house and it was quite peaceful. I held the probes and awaited the shifts to begin taking place. At first I wondered if it was plugged into the wall because nothing "significant" began happening, or so I thought. I sat quietly and peacefully and closed my eyes thinking I might see what I saw earlier in the day — the white shimmering lights. The first thing I noticed was that my entire body got VERY relaxed. Almost as if a nap might be occurring, but I was definitely not sleeping. I noticed vivid purple and blue shifting colors almost as if coming from the top of a church through the stained glass windows. I noticed black shimmering figures amongst the shades of purple/blue. From this point I shifted between states of seeing these black figures with the color purple radiating around them to visions of white clouds, as if there was a tunnel of clouds I would move through. At times the figures I was seeing became more obvious, seeing the "whole body" as a wave form/figure and then at other times just the head, shifting as though it was moving closer. The move through these spaces did not bring any sudden "jolts" as my body remained in an extremely relaxed state and the travel felt light and airy. Then I began to see landscapes, flying above and on the ground at different points. There were plush green landscapes and rivers. At one point I was walking down a road. At another point I was flying above what looked like a large estate in the middle of an English looking landscape and did what felt like a dive into the estate.

At another point I was on a road and came to a road sign that had an upside-down sign on it — with an arrow going the wrong way. It was a sign on top of another yellow road sign on a post in the ground. Traveling just a bit further from this I noticed a vehicle turned upside-down. It looked like a small van. In-between these more vivid images of specific locations I would experience the ongoing movement between traveling through tunnel-like clouds and black figures with purple backgrounds.

At different points there was white/yellow-looking light breaking through on these darker landscapes and noticed a desire to go into the little circles of light when they arose. This entire experience went on for approximately 1 hr/15 min. My body had a most relaxed state the entire time and at times I was not sure I could move, any part of my body. I had to start slow, everything was in slow motion. I did not want to "wake up" but found at one point that it would be necessary and tried to find a spot that seemed natural to re-enter. I felt like I could have remained in that state for several hours and found it rather enjoyable.

My body would feel pain/discomfort in places at times and then the discomfort would shift and be gone. As I tried to awaken myself to the present moment, I took slow steps and moved out feeling more centered, relaxed and with a sense of myself as someone who had more substance, more grounding than previously. I felt excited that I could use this machine everyday if I wanted to develop myself and felt extreme gratitude for this process.

S.P. — Clarks Summit, Pa.

I have been doing a specific chi gong meditation for three years. Now when I wear my Beta-Blocker amulet I find a calm, clear focus which allows me to "see", rather than wander, obsess, imagine, drool or otherwise beta-ramble. The difference is unmistakable.

L.S. — San Francisco, Ca.

The Beacon

Spiral Beta-Blocker

Chapter Fifty-six

Promotional Materials

If you've ever wanted an objective recordable message detector when communicating with the Recently Deceased, Angelic Entities and more, this is the Altar Device for you.

The Spirit Radio is not easy to construct. It is a level of difficulty of 10 on a scale of 10, therefore I build them and ship them already made.

It requires high-level soldering skills and the ability to follow a circuit diagram to successful conclusion, combined with coil winding skills and lots of time on your hands — the project takes a couple of days to complete.

It takes just a few minutes to install it and hook it up to the antenna and ground system. It normally resides on an altar table or workbench.

You merely follow the easy instruction manual on how to operate the Spirit Radio Altar in your own home.

The manual for the operation of the Spirit Radio uses photos, diagrams and text to explain how it works, why it works and how you can learn to use it. It does not explain WHY you want a Spirit Radio; that is the province of your own work path.

You can send for the Spirit Radio today. Location of the vintage parts may take some time. Some parts for the Spirit Radio are so unusual that they must be made by hand, as nothing like them have been available since 1906, when the Marconi Radios were being manufactured. The technology to build the Spirit Radio has been lost for many decades, and the correct values of vintage parts are extremely hard to find, especially when measured in pico-farads and high-impedance devices. Everything today is low-impedance, of course, but modern gear, especially transistor and digital gear, is utterly useless for this purpose.

The completed Spirit Radio includes everything you need except the table on which it will reside.

Claude and I will construct the electronics and wind the coils and build the crystal cages for you and ship you the completed project ready to use.

Here's what you get:

1. Full Instruction Manual on the Use of the Spirit Radio with photos, diagrams and history and theory of the Spirit Radio from its first appearance in 1906 to the present time.
2. 1 Spirit Radio with Hand-Wrapped Tapped-Coil Detector System and air-variable capacitor tank-tuning system with a High-Density Dual-Terminated Quartz-Amethyst Copper-Caged Crystal Antenna Resonating System.
3. 1 Black Tourmaline Cage-Mounted Protection Crystal
4. 1 High-Impedance Earphone
5. 1 Ground Lead connector
6. 1 High-Grade Outdoor Antenna Kit ready for mounting.

Since it takes a couple of days to produce just one of these Spirit Radios, the orders are on a first-come, first served basis. We will construct the units as we are able with our very full workshop, game and book release schedule. We are currently working on six new books in the field of art, work ideas and game development. Vintage Parts are rare. We hope to deliver enough of these to satisfy the demand, but cannot guarantee this. Orders are subject to availability. There are no modern-day substitutes, so when parts run out, we cannot make anymore of these.

BETA-BLOCKER BREAKTHROUGH

Building a working Beta-Blocker amulet takes following a circuit schematic and intense soldering skills and woodcrafting skills, jewelry skills, lapidary skills, and more . . .

The "more" is this:

1. 25 watt soldering iron, $35.
2. roll of electronic Kester resin-core solder, $10
3. special-tipped pliers $85
4. soldering tweezers $55
5. soldering pad, non-asbestos $35-$55
6. small wire clippers $85
7. wire strippers $35
8. roll connecting wire $10
9. roll magnet wire 22AWG $175
10. fine mineral glass pocket-watch crystals $30
11. high-grade black diode $30
12. jewelry-grade sterling or gold-filled bezel $60
13. needle-nosed pliers $85
14. Fancy bail $25
15. Nulling Gear & Meters $4,500
16. QRM gear $450
17. Coil Tanks & Antenna Tuners $300
18. Experience at reading schematics & assembling electronic gear
19. Lots of experience making clean and neat solder joints without pain, and fine silver and goldsmithing techniques . . .

So . . . we thought it better to make them for you, using all the electronic gear and tools we have here at our disposal. During the workshop, attendees with soldering and electronic and woodworking and metalworking skills will be invited to bung up their project, requiring us to step in and repair the mess; Claude and I will still have to BFO and Null the apparatus before you can take it home or we can ship it out to you.

The good news is:

You will end up with a functioning Beta-Blocker for your own use made with vintage and hard-to-find parts.

You don't even have to plug it into a wall socket or install batteries. It derives its power from the differential between the antenna and ground.

Your Beta-Blocker is fully tested before leaving our hands, and is carefully packed and sent by the fastest delivery system you can afford. You get a usable work tool for any art studio, music studio or creative atmosphere; it can even be used successfully in a work cubicle if you wish.

I have completed my experimentation on seven different prototypes of the wearable Beta-Blockers and am currently in production on #3, my favorite and what I consider the best-functioning of the seven basic prototypes.

BFO and Nulling gear is still very expensive, at the $750 range, and you need serious training to learn the skill. I trained for it at the U.S. Army's expense, over a period of almost a year, and haven't yet managed to convey the skill to anyone else, and it isn't for lack of trying. It's damn hard and requires a serious background in electronic communication Ham Radio gear and a rather esoteric knowledge of superhet re-gen theory.

The cost of assembly is too much in my opinion, with all the tools you would need, which I take for granted, because I've had them all my adult life . . . In addition, high-level soldering skills are essential unless you want to have a permanent scar on your hand or face.

Spirit Radios are still under experimentation. I have one working well but am not yet satisfied with some issues. I will be making them rather than sending out kits, for the same reasons as above; the tools and testing gear are far too expensive for the average person. I will give classes in crystal radio construction at our new Brane-Power training center in Grass Valley.

HUGE BREAKTHROUGH!!! BLACK DIODE MAKES POWER-PACKED BETA-BLOCK AMULET!!!

A friend in the amateur radio network sent me an incredible BLACK DIODE which he claimed would outperform virtually ANY crystal detector in any circuit. I duly worked up a test circuit board with Fahnestock clips and went to work on it starting at about two o'clock this a.m. —

Here it is almost five a.m. and I've just finished testing dozens of different elements in my crystal radio circuits and . . . true to Scott's word, the BLACK DIODE was incredibly powerful, at least 100 times the power of the 1N34A I've been using!

Problem is, they're mighty rare pre-war items and I can't get many of them and they're no longer made, haven't been made for many years now, because they're uniquely useful in crystal radios, and who but us parallel universe explorers are interested in the subject anymore???

One really fun item, which we got, at least part of it, was that when I tested the BLACK DIODE circuit with the powerful BIG BLUE CAPACITOR, the program we landed on was a talk show on the subject of parallel universes, psychic self-defense and extra-dimensional travel using crystal radio technology!!!

The most fun part was that one of the guests on the show was speculating — and this is on the video — postulating a universe where THE NORTH WON THE CIVIL WAR and HITLER LOST!

In other words, we tuned right into an alternate universe on the radio!!! Most folks might not catch the drift, but it was clear that they had lived through a number of historical events which are very different from the ones we think we know "actually happened", but which are not the same from one world to the next.

I lost no time in making a dollar-sized amulet out of the BLACK DIODE and put a silver bezel around it. I'll be using MINERAL GLASS on both sides, to improve the silicon levels, and of course there's the GREEN ANTENNA COIL to top it all off.

If you want one of my BLACK DIODE Beta-Blocker Amulets, better order NOW while supplies last. I cannot guarantee that Scott and I will find many more of these components, although he promised to try to hunt some up from friends/collectors/ham radio hobbyists who might have some around.

Don't even ask about a smaller amulet — I was lucky to be able to cram these much larger circuit components into the dollar-sized amulet, and it definitely needs the larger coil.

I'll be testing this amulet with my Deep Delta Parallel Universe Explorer Induction CDs today. I'll let you know the results as soon as I see them myself. I'll publish the reports from my guinea pig testers the minute I get them in my hands.

Let me know what you think of my official Beta-Blocker site, http://www.yoyodyneindustries.com.

BEACON MK II

Having spent the better part of a week thinking about how to go about it, I dedicated the major part of my night workshop time — the hours when I generally produce items for shipping to fulfill orders for things — working on a new circuit and casing for the Beacon, because I realized that there was a need for a smaller unit for those who want to use them in coffee houses and small apartments, offices and/or cubicle workstations in the corporate sector.

It turns out to be the same powerhouse as the Beacon MK I and to have pretty much the same operating characteristics. With the exception of size, it's basically the same unit as the MK I, but it's smaller, much smaller.

In a Brane-Power Center, you'll probably want the larger model, but there are many applications and situational conditions where a smaller model might come in real handy, so I spent the time putting one together. It's online on the yoyodyneindustries.com site for your perusal.

If you asked me which one you should get, I'd respond that it's all a question of where you intend to use it and whether it needs to be extremely portable or not.

If you're planning a large center with lots of usage, I'd definitely recommend the MK I. On the other hand, if it's just for you and you're going to want to stash it away in a desk drawer, choose the MK II. As I said, they both have equal power and are both easy to use.

You can also select a variety of sizes of grips to fit any hand. I'm making some now for a giant-sized voyager who stands 6’5”, and another for a tiny bird of a voyager whose hand size is about that of a large doll.

How do you determine your grip size? Send me a photo of your hand open and another with hand closed, next to an item of known size, such as a quarter, half-dollar or silver dollar, or a sharpie pen . . . anything I can use to estimate the size correctly.

Another way to measure your grip is to get hold of a roll of pennies, nickels, dimes, quarters and half dollars (half dollars might be tough) and see which one fits your hand the best. That will really help me to customize them to your exact fit.

Another customization item — you can order your Beacon finished in a variety of stains. You can specify your preference for mahogany, maple, oak, walnut or greenwood.

In case you didn't know this, I personally make all the Beacons and in fact all the models of Beta-Blockers, which includes skills in woodworking, soldering, jewelry making, circuit schematics and ham radio skills. I leave all the math and explanations of WHY things work to Claude. I'm the theoretical physicist who blunders into solutions, and I work well within those limits.

Let me know as soon as possible which type of Beacon you want, for three reasons:

1. First come, first served, and there's a long, long list now of folks who want one or another one. Some folks have ordered several for their Brane-Power Centers in their hometowns.
2. Parts are hard to find.
3. Time is hard to parcel out — I'm working at these an average of 12-14 hours a day, but I do have other things I need to do and I'll have to start slowing down a bit on Beacon and Black Diode production. If it goes into the high hundreds or thousands of requests for these, I'm afraid I'll have no choice but to outsource the work to China, Pakistan, India or Vancouver — several folks have offered to help out and I will at some point be forced to accept those offers — I need time to work on GODD levels and at least a couple hours a day to run intels and do Baal runs!!!

Promotional Release

This combination of radio wave technology, beat-frequency oscillation, harmonic resonance, waveform reduction and electronic biofeedback of the salt-water electrolytic biocomputer (commonly referred to as the "human being") is the registered invention of E.J. Gold, early biofeedback pioneer and author of the 1964 classic, "Alternate Worlds Beyond the Galaxy".

Beta-Blockers use radio waves in nulled beat-frequency oscillation to produce a reduction wave at the harmonic frequency best suited for deep meditation, called the "Schumann Harmonic" at 7.8 Hz. without training or effort.

The Beta-Blocker is a new creative use of the early pioneer crystal radio technology. It derives its power from the radio waves themselves and needs no battery or electrical hookup to operate. Your Beta-Blocker is a highly modified (hacked) working crystal radio.

During the early days of radio, nothing was known of brainwave technology and biofeedback. By the days of biofeedback, everything about crystal radio technology had been lost. This revival of pioneer crystal technology combined with brainwave biofeedback is the result of an accidental discovery by 1960s biofeedback pioneer E.J. Gold.

You can wear your Beta-Blocker Amulet all day and even at night, or place it bedside if you prefer. It allows you to instantly achieve the deep meditation of a Zen Master without training, totally without effort.

Nothing to practice, nothing to join, just pure simple deep meditation from the first moment you put it on!

Your Beta-Blocker was handmade in the United States and is encased in a .925 sterling silver or 12k gold-filled locket. They are also available in a solid 14k gold locket if you prefer — price of gold changes daily, so call for quote.

How to use your Beta-Blocker: Don't worry, be happy!!! You will find new bliss when you allow the Beta-Blocker to do its work to make your day happier and better! The benefits of deep meditation are many, the most important of which is that when you feel good, the whole universe is good!

3 SECONDS TO INFINITY

That's right. Within 3 seconds on the Beta-Blocker you'll be feeling an instant and automatic expansion into infinity, the world of deep meditation.

You'll experience what it feels like to be a Zen Master without the years and years of extensive practice, all within the first 3 seconds!

It's fast, it's fun and it's guaranteed to take you there without drugs, hours of pretzel-time in strange positions, without bizarre breathing techniques and without any effort.

It's totally effortless, and no meditation skills are required to achieve instant and effective stress-busting levels of deep meditation and restorative power.

Handle life better with greater confidence. Lower stress with Alpha-Theta Brainwave Power. Achieve success and strengthen your self-esteem with Brane-Power.

Gather your Parallel World Selves into a powerful cluster of charismatic unity!

Inner Peace, Harmony, Creativity, Self-Confidence, Awareness, Awakening Cosmic Consciousness are yours instantly without effort, without training. All within 3 seconds of your time!

Combat anxiety, fears, depression, indecision, fear of success, epic fail, confusion, procrastination, workplace stress, substance abuse with deep levels of instant meditation.

No training! No Belief System! Nothing to Join! Just 3 seconds of your time and you'll feel instantly connected with the Cosmos, infinitely grounded in spirit!

Try Brane-Power Beta-Blocking absolutely FREE! You'll be glad you did!

HELLGAMER BLACK TOURMALINE BEACON

You might have been wondering what you've seen recently on gorebaggtv — it's a Black Tourmaline Beacon. Not only are the grippers made with giant black tourmaline crystals, but the whole circuit board is different, very different.

What's it for? Well, after the first few levels of PUP connections, you'll find yourself wading through a number of very dangerous spaces with equally dangerous entities waiting to more or less gobble you up.

Without serious protection, you'd be open to all sorts of psychic invasions, so we switch from crystal quartz to crystal black tourmaline, with corresponding switches in frequency and modulation.

I've just spent $4000 obtaining the necessary black tourmaline crystals for our center; we'll need them in our advanced training.

While waiting for the black tourmaline crystals to arrive, I haven't wasted any time — I've been crafting up the Hellgamer Black Tourmaline Beacon which utilizes their natural crystal wave-filtering frequencies. The crystals arrived yesterday afternoon — Iven ran them up to me from the receiving area in shipping and I quickly modeled some grippers for them.

Because of the strange and peculiar shapes and sizes of black tourmaline crystals, I had to find new and different ways to accommodate them with the heavy copper spirals, which necessitated employing my silversmithing and coppersmithing skills, which I got from my PUP connections with L315a and K313c.

Wire-bending is difficult at best and, without a jig or form to work against, it's doubly hard. In addition, the copper wire is 10 and 8 gauge, which is mighty thick and intractable stuff.

Undaunted by the odds against success, I plowed on with the project and by dinner time at 6:15 (dinner is normally at 5:30, but Marvette was on cook duty last night, and she's a perfectionist, which means dinner on her shift is generally late) I had the gripper concept down pretty good.

I made about 30, maybe 35 grippers with black tourmaline, and I'm happy with the result. I grabbed a pair of tourmaline grippers this morning and went at it with the new Hellgamer Beacon, which by the way is the same wood case, but stained bluish-green and polished with a power buffer.

The net result was a very good and rather exciting adventure in the darker zones of PUP retrieval.

In addition to all that, I've been experimenting with group dimensional shifting where we all move together from one reality stream to another. It has been very successful — maybe you've noticed a profound shift in levels of prosperity and grokking. To Grok means to dig it, deep levels of understanding . . . comes from Robert Heinlein's "Stranger in a Strange Land" which, if you haven't yet read it, would be a good addition to your background information regarding interdimensional . . . well, grokking.

Work on the training center is coming along nicely. We'll have a great pilot program set up for you to work with as you set up your very own hometown Brane-Power Centers.

As I've mentioned before, I'm very willing to train trainers — I can't go everywhere to do this, so you'll have to come here to my training center in Grass Valley, California, at 207 Neal Street. It's small, it's very small, but we don't need much space to work with, just enough for about 20 folks to train in. If we get more folks at any one time, we have enough training space around the corner for about 50-60 more.

If you want personal training, you can get it here at my new training center. I've trained several trainers to work with trainees on PUP retrieval.

If you want me to work with you on your personal training, I'm available for that on a limited basis — limited because I have art commissions, book deadlines, experiments to perform with new Beta-Blocker devices, and of course online gaming which takes precedence over all other agendas.

One of the most popular concepts regarding the Beta-Blockers is that of the "oxygen bar" or "cat-napping" kiosks. Check this out at your local mall or downtown area . . . you'll typically find at least one of each in your hometown, no matter how small it might be.

The idea is that your client comes into your Brane-Power Center and sits with the crystal grippers patched into the Beacon for about three minutes. That's all it takes to pump up the spiritual energy levels. Then they can continue to hold the crystals for a while or not, as they choose. Most of the folks I've worked with so far have preferred to paint or draw with that energy and then get back to work. What a terrific idea for lunchbreaks for those poor souls trapped in cubicles all day!

So . . . it's back to the electronics bench for me . . . I've got an idea for a wearable Beta-Blocker amulet that's smaller than a dime! Will it work? Of course it will. The only questions are how small I can make it and how much more it will cost to miniaturize it. I'll post the results on yoyodyneindustries.com when I get it working. My plan is to put the device into a 14k solid gold bezel with hand-ground mineral glass parabolic lenses. Should be a doozy, if I get lucky with the circuit design.

Oh, yes . . . almost forgot to mention it — several folks are experimenting with my black tourmaline, amethyst, citrine and fuchsite hand-held spirals. They're like the grippers except you don't have to plug them into anything . . . I've made them so they work across the body through the chest — the thorax area.

Black Tourmaline Grippers are available now from my online outlet at http://www.yoyodyneindustries.com — the things are pricey-ish, but not when you consider the massive size of tourmaline crystals and the amount of work that goes into the spiral grippers.

The Black Tourmaline is not the ordinary variety — you won't find this kind of material in your local rock shop, same as the phantom quartz, copal and blue chalcedony that I'm obtaining for my Stand-Alone Grippers.

What are they good for? What do they do? You tell me! They're meant to be used in a process of discovery! I'll be interested to read your logbook entries on what you've found out about them and their many possible uses!

MASTERS COURSE

If you own a Beacon Strapper and you use it to go through Level III PUP Training, you will be eligible at that time to take the Master Trainer Course, after which you will be able to give Upper Level training to others, using your own Beacon and Strappers. At that time you will be able to deliver up through Level III Training.

If you start a Brane-Power Center, you are eligible to administer Level I PUP Training immediately, using the induction disks that I have provided, which are the Beacon Strapper CDs — there are 30 of them in the complete set. You can immediately upon receipt of the Beacon Strapper administer the Level I Training!

I ask $15 per training, which is 10 minutes on the Beacon, plus 5 minutes before and about 5 minutes after for orientation, questions, etc. So it comes to about 20 minutes per person in the first few trainings. After the fourth or fifth training, questions and confusions seem to settle down a bit.

If I gave the same training in New York City, I'd probably ask $50, because the cost of living and delivering training is so much higher. Keep in mind that your training center is NOT owned by or part of the Institute — it's your own thing. I merely suggest that you keep the donations to a reasonable level so folks in your hometown can avail themselves of the training without pain.

I get 8 to 10 people a day in the center for training, and that's a nice number for me to deal with on a daily basis. I suppose I could handle 15 to 20 without too much stress, but above that, I'd need someone to help with the administration.

Don't forget that there is a bit of paperwork involved. The logbook must be filled out for each training event, and a copy should be typed up and sent via email to my office for evaluation and inclusion in my Big Logbook. I'll select the choicest entries and put them in my book about the training, which will be out, with luck, sometime by the end of 2009.

Keep in mind that this spiritual training is NOT intended for the diagnosis, prevention or treatment of ANY disease. The Beacon is NOT a medical device nor does it have any medical purpose or use. It is strictly a deep meditation device for use in the spiritual path as outlined in the ABD and other Institute resource and study materials.

I assume no liability for the use or abuse of any spiritual device or instruction delivered by me for the greater good. I wish this to be used for the benefit of all beings everywhere, and I hope you do, too.

EL BRUJO

This special high-powered amulet is not for casual wear. It is intended for the Brujeria's curanderismo, which is to say without the use of spells or incantations, using instead psychic-intuitive powers for voyaging and shapeshifting, and for the Brujo's true work as a curandero for the wandering soul. Soul-Retrieval is a powerful purpose for the Brujeria, and El Brujo is particularly useful for shapeshifting into an owl, coyote or cat and is especially useful against malevolent spirits.

El Brujo takes its name both from the shamanic voyager and from the small Moche Huaca Cao Viejo pyramid, whose courtyard was used for ancient transformational and interdimensional ceremonies.

Tattooed mummies, ceramic vessels and wall paintings found at El Brujo depict parallel universes and wormhole voyaging and predict the coming of the trans-dimensional device known as the Beacon.

WHAT IS THE PLAN?

Wear your Circle of Peace Beta-Blocker all day every day amongst as many people as you can.

Influence people around you by bringing Peace and Harmony everywhere you go.

When people ask you how they can help, tell them about the Circle of Peace Beta-Blocker.

Give Circle of Peace Beta-Blockers as gifts at Christmas, Chanukah, Birthdays, Graduations, Weddings, Confirmations, Baptisms — when gift-giving is appropriate, the Circle of Peace Beta-Blocker is the gift that keeps on giving.

Sell Circle of Peace Beta-Blockers. If you have a gift shop, boutique or small business of any kind, put a Circle of Peace Beta-Blocker display in your workspace. One couple who operated a CPA firm in Denver actually retired from accounting and went full-time with Circle of Peace Beta-Blockers because their amulet income exceeded their accounting firm's income!

Go door-to-door with Circle of Peace Beta-Blockers.

Obtain a street-vendor's license and sell Circle of Peace Beta-Blockers to passers-by. You will be surprised how many people will love to wear them not only for what they do, but for how cool they look! It's a great piece of jewelry and attracts the eye as well as pacifies the neighborhood.

Give Circle of Peace Beta-Blockers to powerful people — politicians, business magnates, wealthy and influential power-possessors, celebrities, movie and television stars, divas and bloggers! Let them tell their fans the story!

Get onto talk shows, blogs, forums, internet games, anywhere you can reach people, and tell them about the Circle of Peace Beta-Blockers and explain to them how they can help. Don't just Visualize World Peace — Make It So!

PROSPERITY CHARM

When I gave my mother Eve my lottery ticket and told her to wear a Prosperity Charm to the California Big Spin, she laughed. But to humor me, she wore the Prosperity Charm Amulet, along with her high-fashion ensemble.

She laughed at the idea that the charm was going to have any effect. After all, the Big Spin was now up to 15 million dollars, and nobody had hit it for 14 weeks straight. Some people thought it was jinxed.

But then, having waited until all the other contestants had spun the wheel, to little or no effect, she got up on the stage and gave it a mighty whirl. The wheel went round and round and stopped right on the Big Number — $15,220,000.00. She couldn't believe it and in fact did not understand until about ten minutes later that she'd hit The Big One.

At the press conference she referred jokingly to my Prosperity Charm Amulet and said that it might have been the cause of her luck. The press people also laughed at the idea.

But then, when she wore the Prosperity Charm Amulet a few weeks later and once again hit the lottery, this time for $50,000.00, she began to be a believer. Still, she refused to quite go all the way with it.

When did she become a total believer in the Prosperity Charm? When she hit the lottery for the third time in a row.

That was the same week she won the Bill Harrah Blackjack Tournament in Lake Tahoe, which, with the Prosperity Charm Amulet firmly in hand, she won two more times, and was featured on a poster at Harrah's Club for several years to come.

After the fourth lottery win, she stopped laughing and kept the amulet with her always.

She had been diagnosed with cancer just a few weeks before the Big Spin lottery win in 1986. The prognosis was not good. She had at most only a few months to live.

Imagine the doctors' surprise when she outlived several of them, until her passing in May of 2009, almost a quarter of a century past the deadline they'd given her to live.

The Prosperity Amulet? Possibly. But one thing is undeniable.

She was the luckiest woman alive, and that definitely was the Prosperity Charm Amulet at work.

Will it do the same for you? I believe that it can change your luck for the better. Whether you win the lottery for 15 million dollars or you find your soul mate, the perfect job or the perfect home, I believe you will see a profound change in your luck, and definitely for the better.

Is this magic? I think it is, but some folks say that today's magic is tomorrow's science. If you'd shown someone a car or a tv set or a cell phone back in the Middle Ages, you'd have been burned at the stake.

I believe that someday someone will have the answer, will know why it is that some powerful charms or amulets seem to bring luck to the wearer. Until then, call it magic, call it fringe science, call it nonsense. But before you laugh it off, do what my mother did. Give it a try. You'll be happy you did.

In this Parallel World, the swine flu wiped out millions — the military had to take control of what was left.

Chapter Fifty-seven

UFO CHRISTMAS CAROL

I'm dreaming of a parallel dimension,
Just like the one I'm in right now,
Where the trees are purple,
The sky is yellow,
And I'm a multi-colored cow.

(chorus)

Parallel Universe is the world next door.

Equip yourself with PUPs before you transfer to other worlds . . .

Contact Information

Dear Reader:

To find out more about Parallel Worlds visit brane-power.com or contact us directly at:

Gateways Books
P.O. Box 370
Nevada City, CA. 95959-0370

Phone: (800) 869-0658
(530) 271-2239
Fax: (530) 272-0364
Websites: www.brane-power.com
www.yoyodyneindustries.com
www.gatewaysbooksandtapes.com